Mathematics

Numbers 1-100

Copyright © 2021 by A Devi Thangamaniam.

All right reserved. No part of this publication may be reproduced, distributed, or transmitted in any form or by any means, including photocopying, recording, or other electronic or mechanical methods, without the prior written permission of the author, except in the case of brief quotations embodied in critical reviews and certain other non-commercial uses permitted by copyright law.

Information: MiLu Children's Educational Source. www.my-willing.com
ISBN: 979 - 8 - 88525 - 436 - 6

1	2	3	4	5	6	7	8	9	10
11	12	13	14	15	16	17	18	19	20
21	22	23	24	25	26	27	28	29	30
31	32	33	34	35	36	37	38	39	40
41	42							43	44
45	46							47	48
49	50							51	52
53	54							55	56
57	58							59	60
61	62							63	64
65	66							67	68
69	70	71	72	73	74	75	76	77	78
79	80	81	82	83	84	85	86	87	88
89	90	91	92	93	94	95	96	97	98
99	100								

In the early years, numeracy is a good foundation and creates a solid mathematical concept. There is a step by step wide variety of practices in numbers 1-100 for young children. Take practice and gain children's intellectual skills.

Have fun with numbers 1-100

Group the pictures for numbers

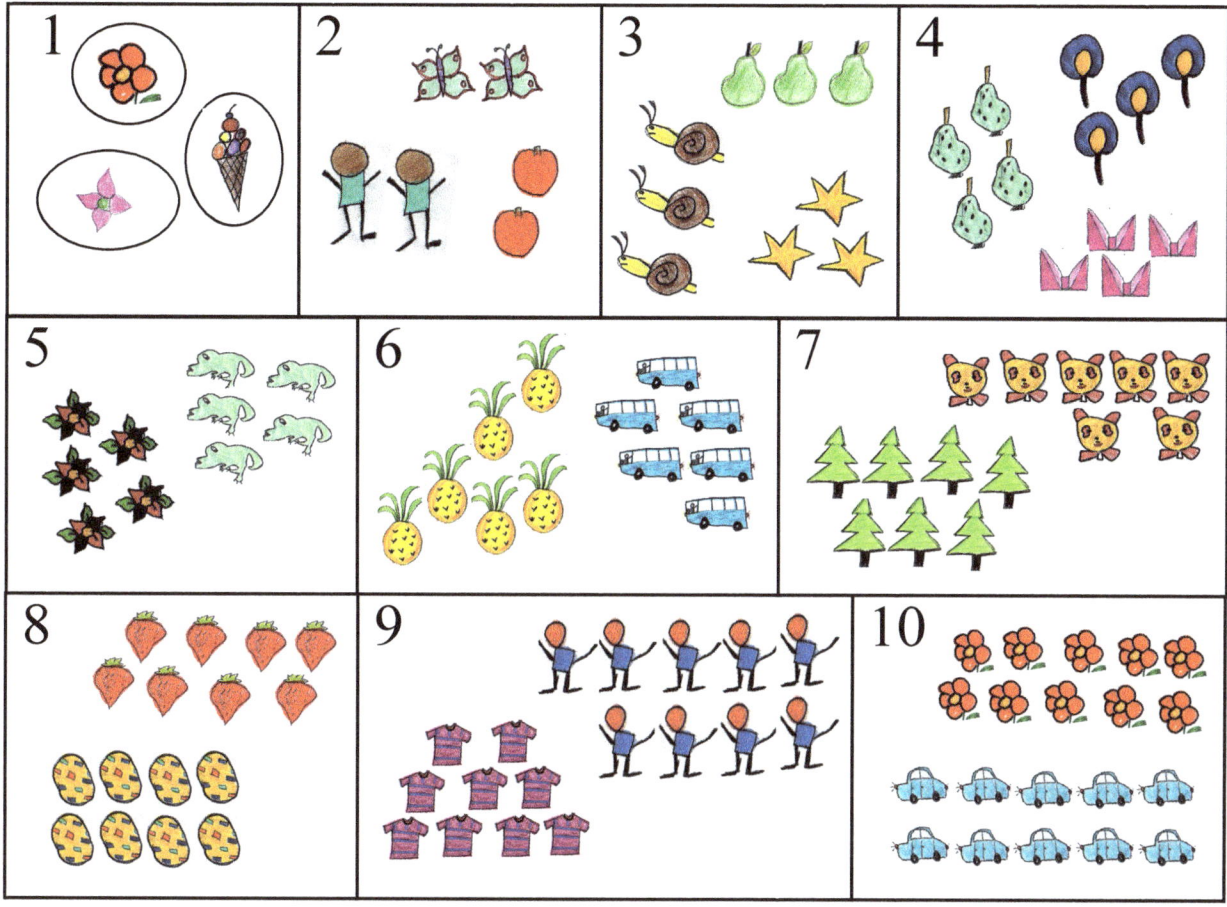

Colour and count the pictures and write the numbers

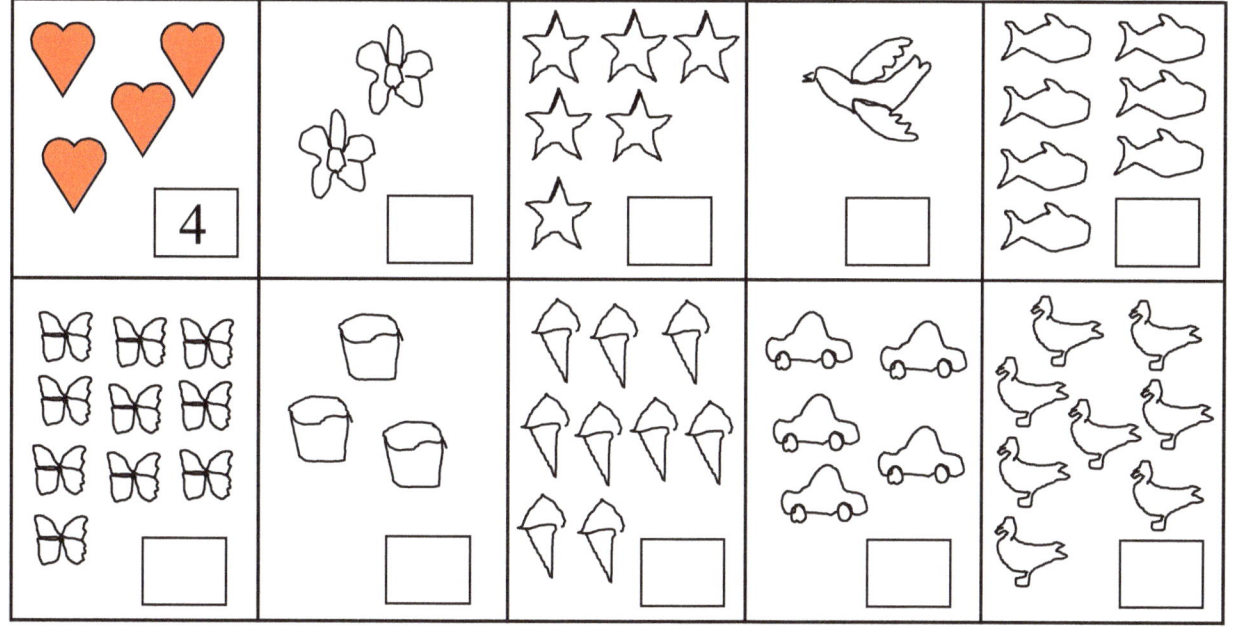

Colour these squares, and connect with numbers

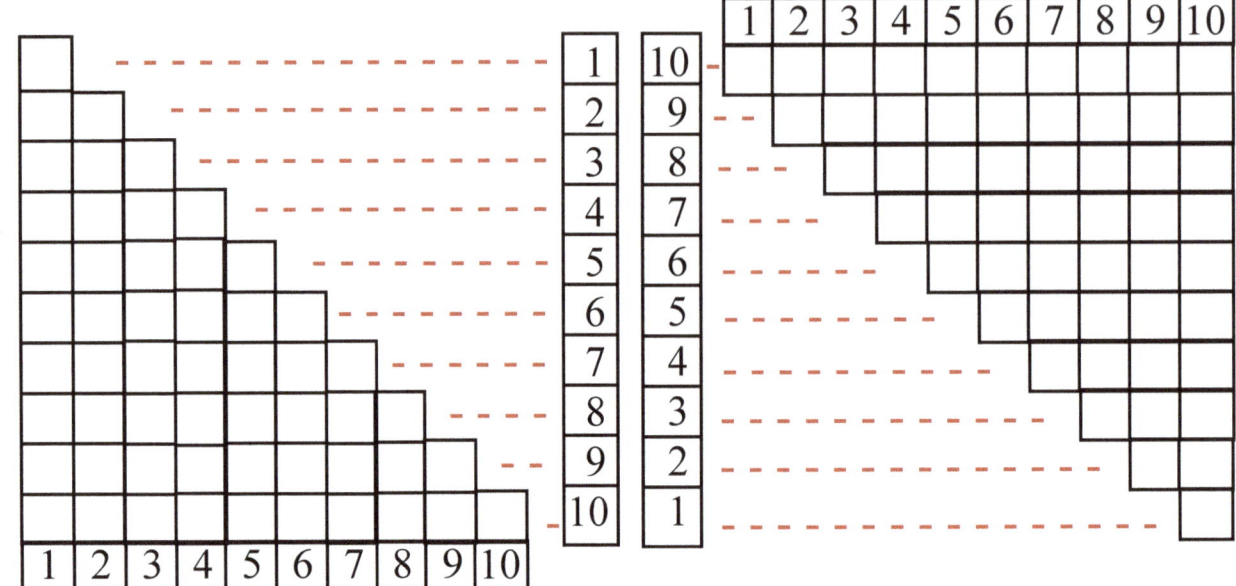

Trace the numbers 1-10 and 10-1

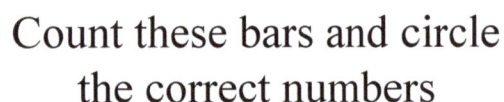

Count these bars and circle the correct numbers

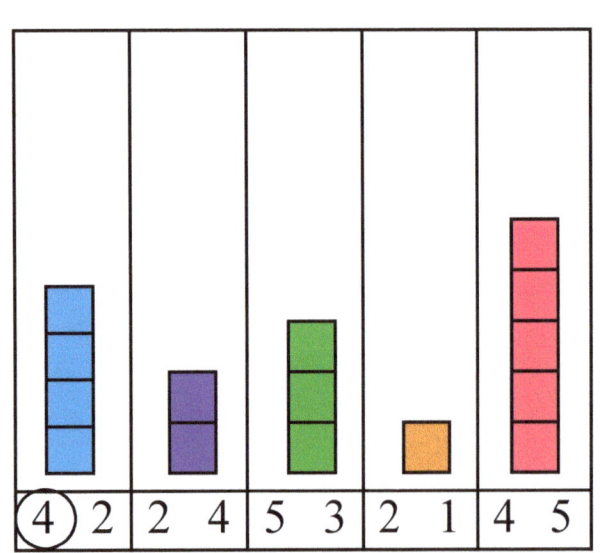

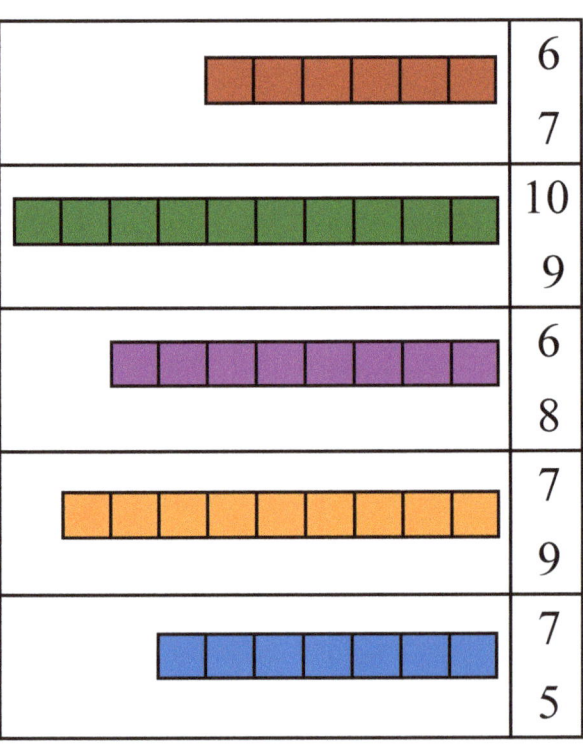

Colour and compare the bars and circle the biggest number

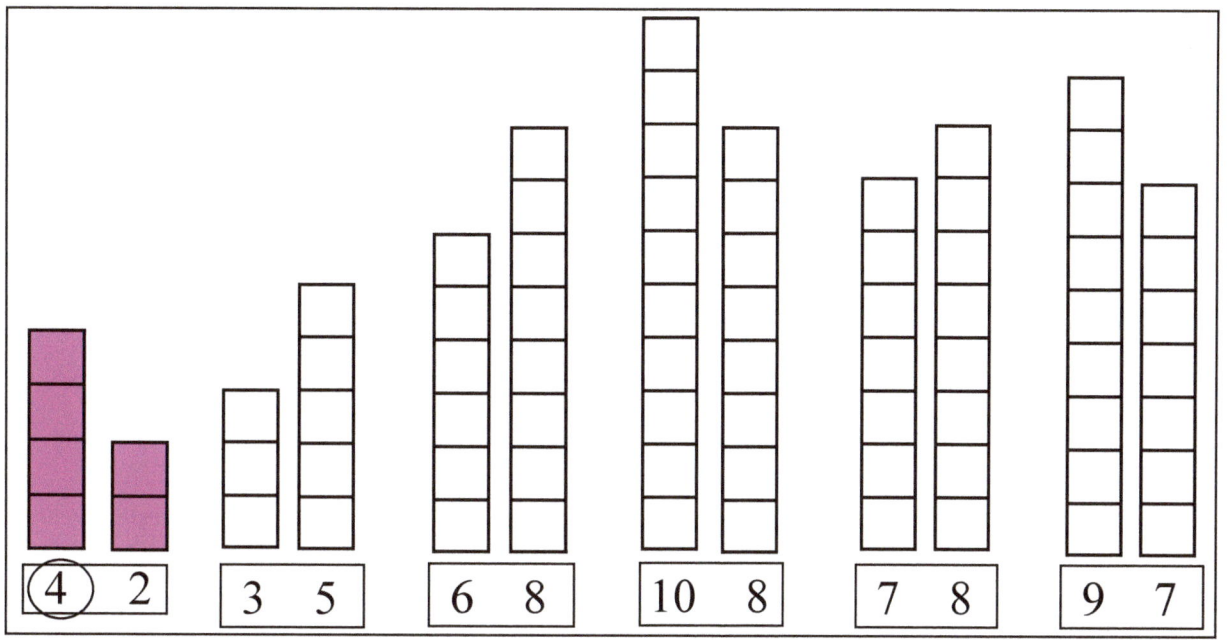

Compare and circle the biggest numbers

5	1	3
8	7	4
4	6	3
3	1	2

7	6	3
2	3	4
6	3	5
6	9	7

10	8	9
4	2	1
5	6	7
9	4	3

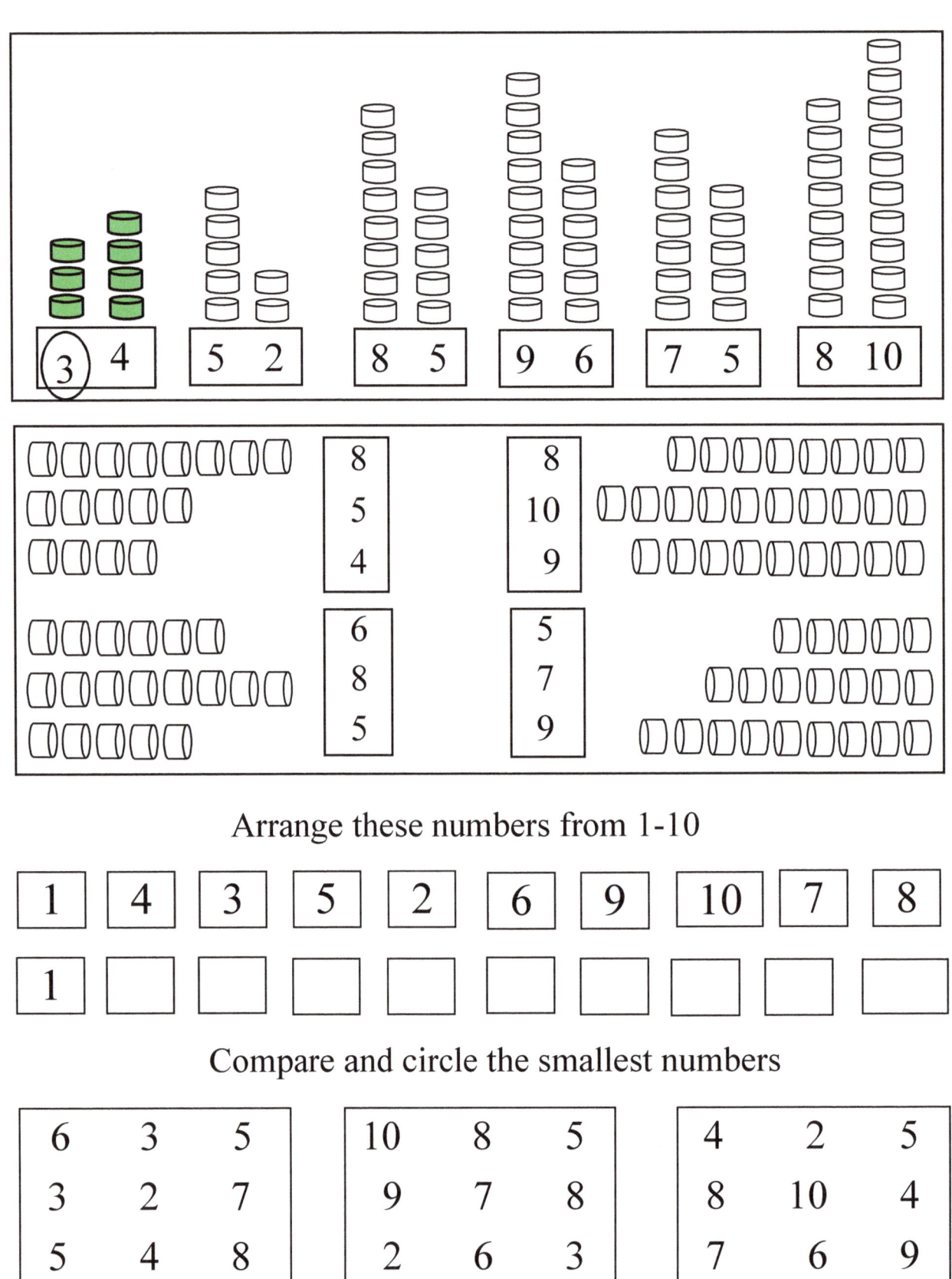

Make a group with numbers and their words

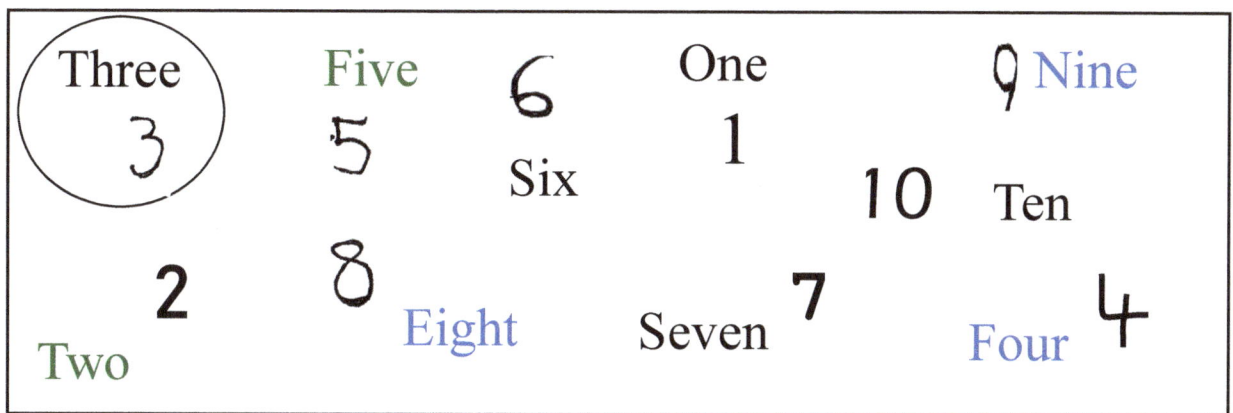

Match the numbers with their words

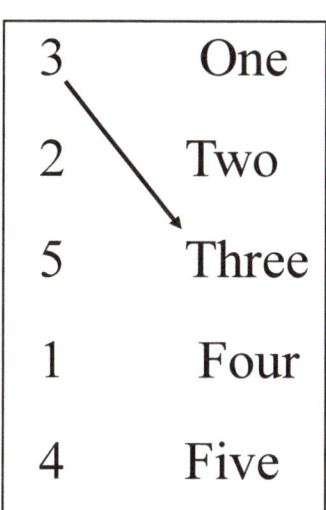

Trace the numbers and their words

1	2	3	4	5
One	Two	Three	Four	Five

6	7	8	9	10
Six	Seven	Eight	Nine	Ten

Find the digit words and circle them

s	e	v	e	n	
i				i	t
x		t	e	n	h
		o	n	e	r
f	o	u	r		e
i	t	w	o		e
v					
e	i	g	h	t	

7 Six
9 Seven
6 Eight
10 Nine
8 Ten

Match with same amount of the dominos

Match with the same numbers

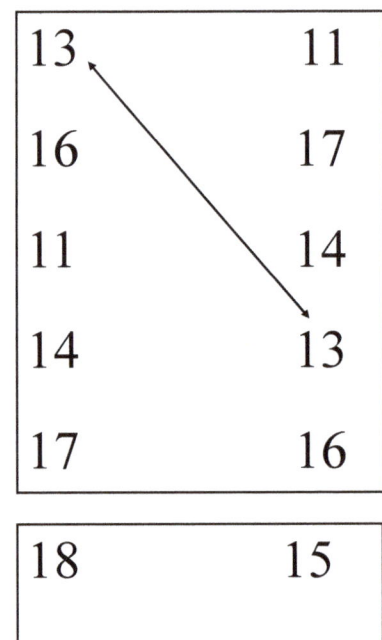

13	11
16	17
11	14
14	13
17	16

18	15
15	19
12	12
19	18
20	20

Arrange these numbers from big to small

12	14	15	13

16	15	14	13

18	17	19	16

19	20	17	18

Make a group with same numbers

12 12 12 13 13 13 11 11 11 15 15 15 14 14 14

18 18 18 16 16 16 19 19 19 17 17 17 20 20 20

Write the numbers which are dotted on the number line

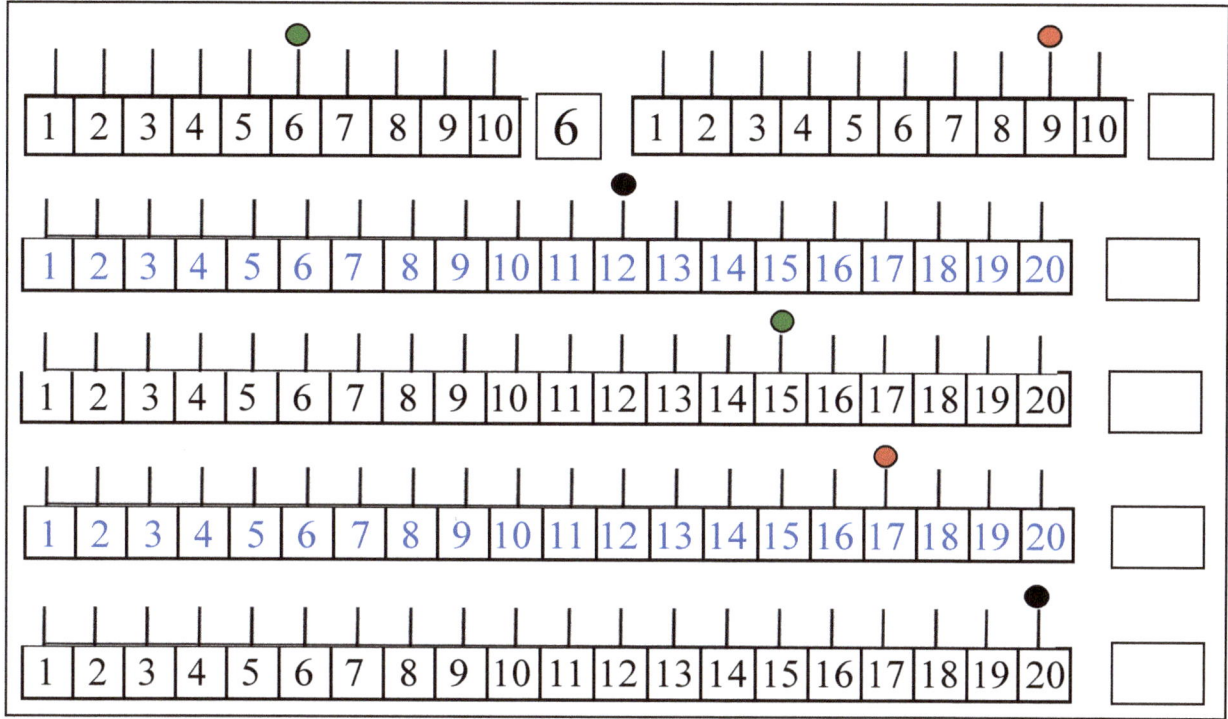

Read the numbers and mark dot on the number line

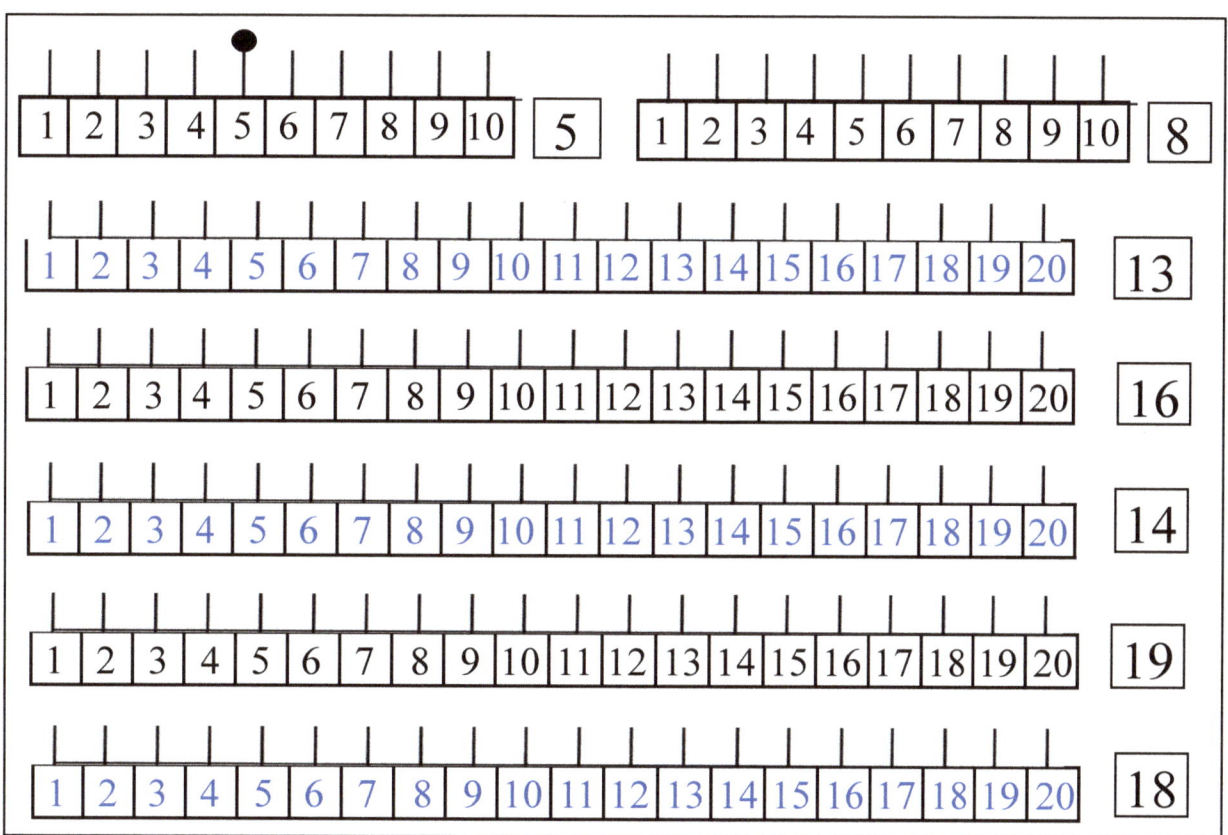

Make a set into two equal groups

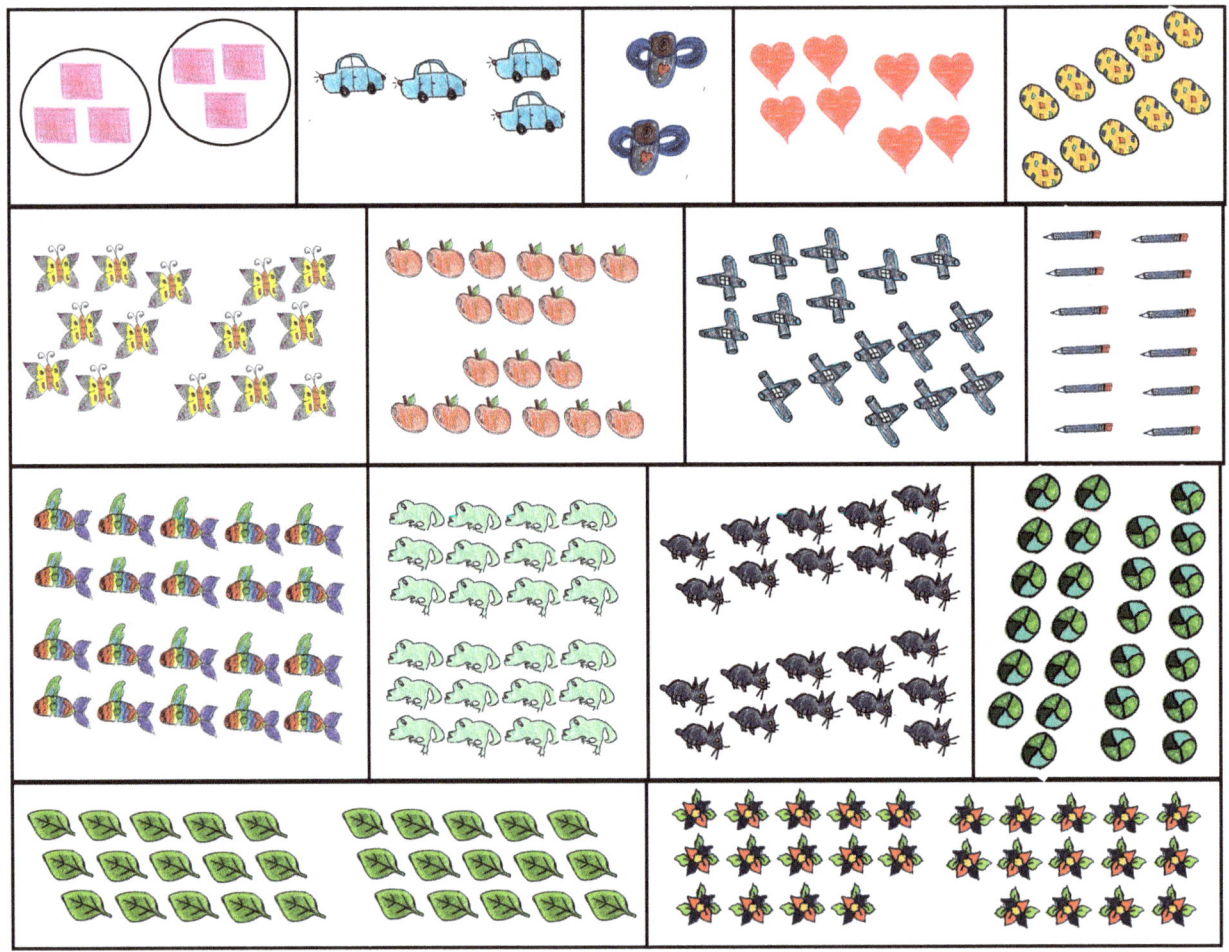

Arrange these numbers from small to big

16	15	13	14

12	11	13	14

20	18	17	19

17	15	18	16

Connect the digit words in order 11 - 20

11 Eleven   15 Fifteen

13 Thirteen

12 Twelve   14 Fourteen

18 Eighteen   16 Sixteen

17 Seventeen

19 Nineteen   20 Twenty

Colour the pictures and group them with numbers and words

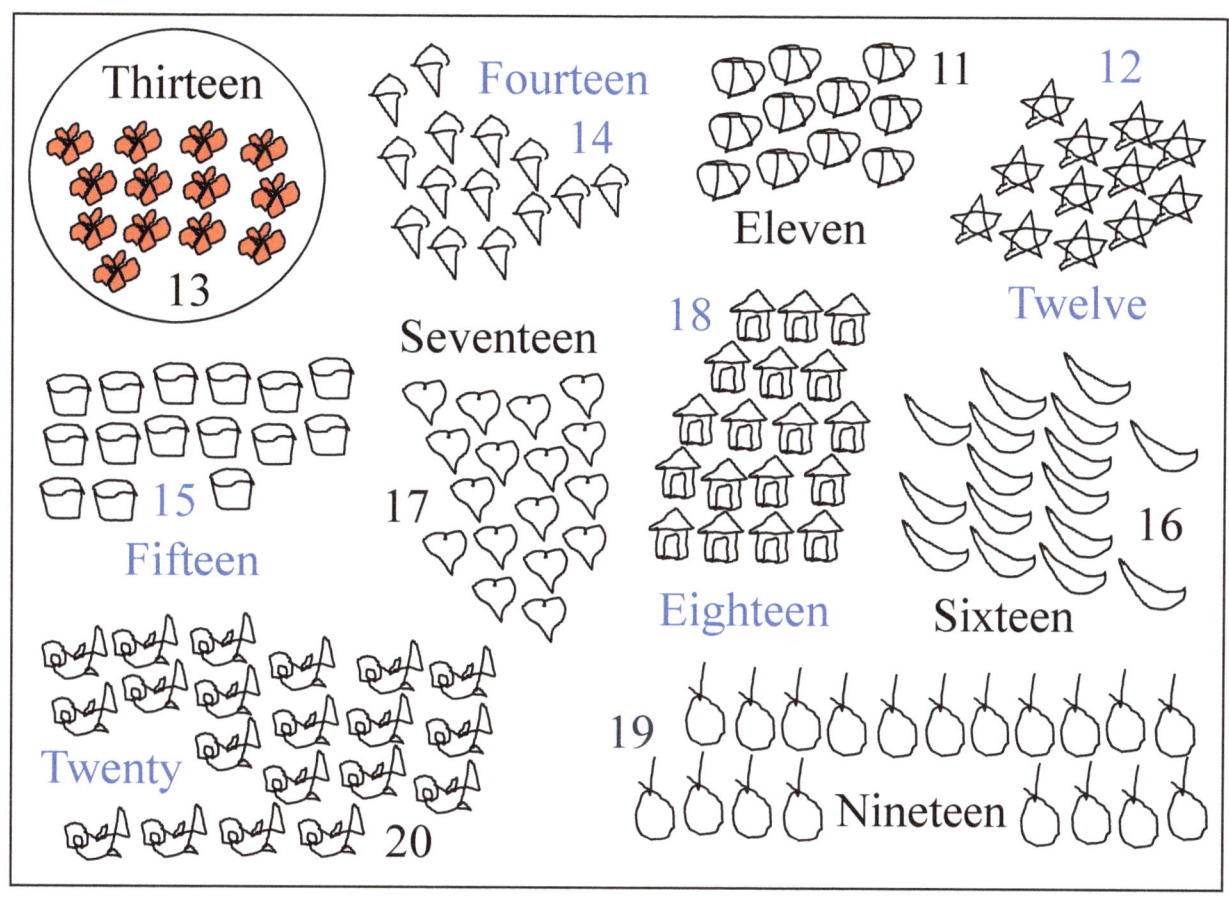

Match the numbers with their words

12	Eleven
14	Fifteen
11	Twelve
13	Thirteen
15	Fourteen

Nineteen	18
Sixteen	20
Eighteen	17
Seventeen	19
Twenty	16

Fill in the missing letters and make digit words

11	12	13	14	15
Ele _ en	Tw _ lve	Thirtee _	Fourte _ n	Fi _ teen
16	17	18	19	20
Sixt _ _ n	S _ v _ nteen	E _ gh _ een	Ni _ e _ een	Tw _ n _ y

Fill the missing numbers in ascending and descending order

Fill the missing numbers with pattern in the horizontal way

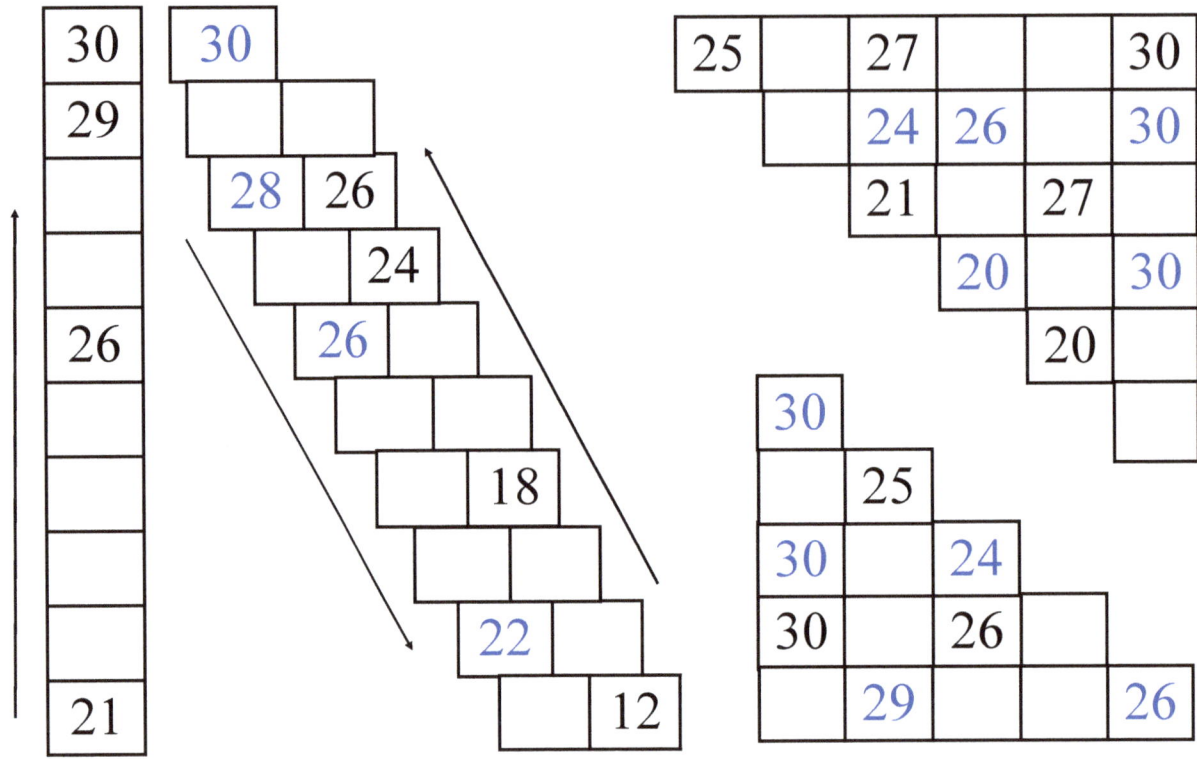

Trace the numbers 21 - 30 and 30 - 21

21	22	23	24	25	26	27	28	29	30
30	29	28	27	26	25	24	23	22	21

Match with the same numbers 21–30

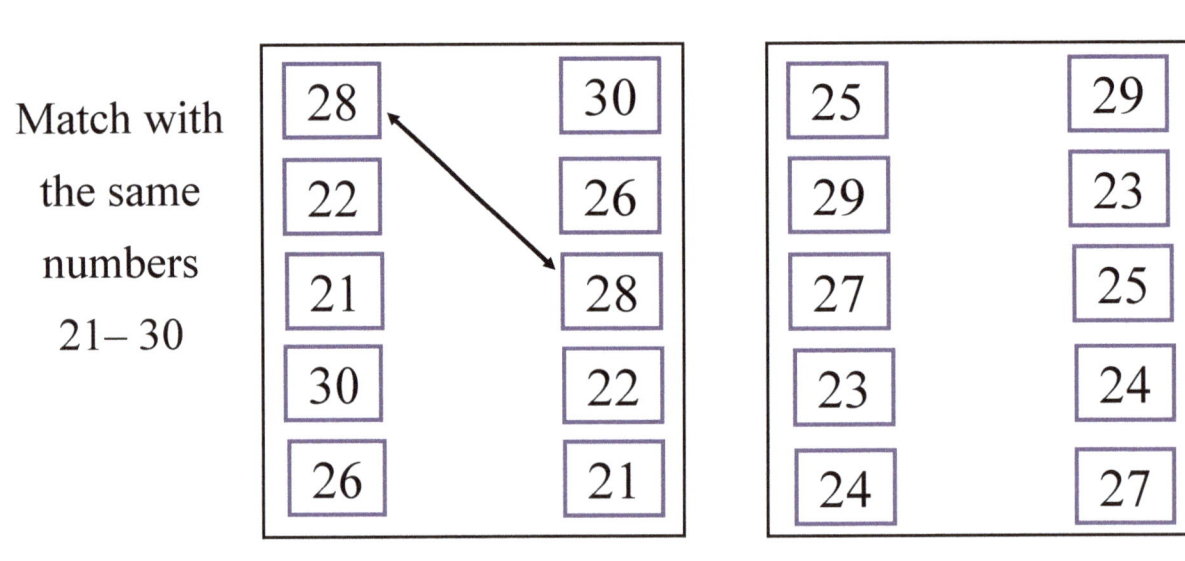

Count with fingers and write the numbers in the boxes

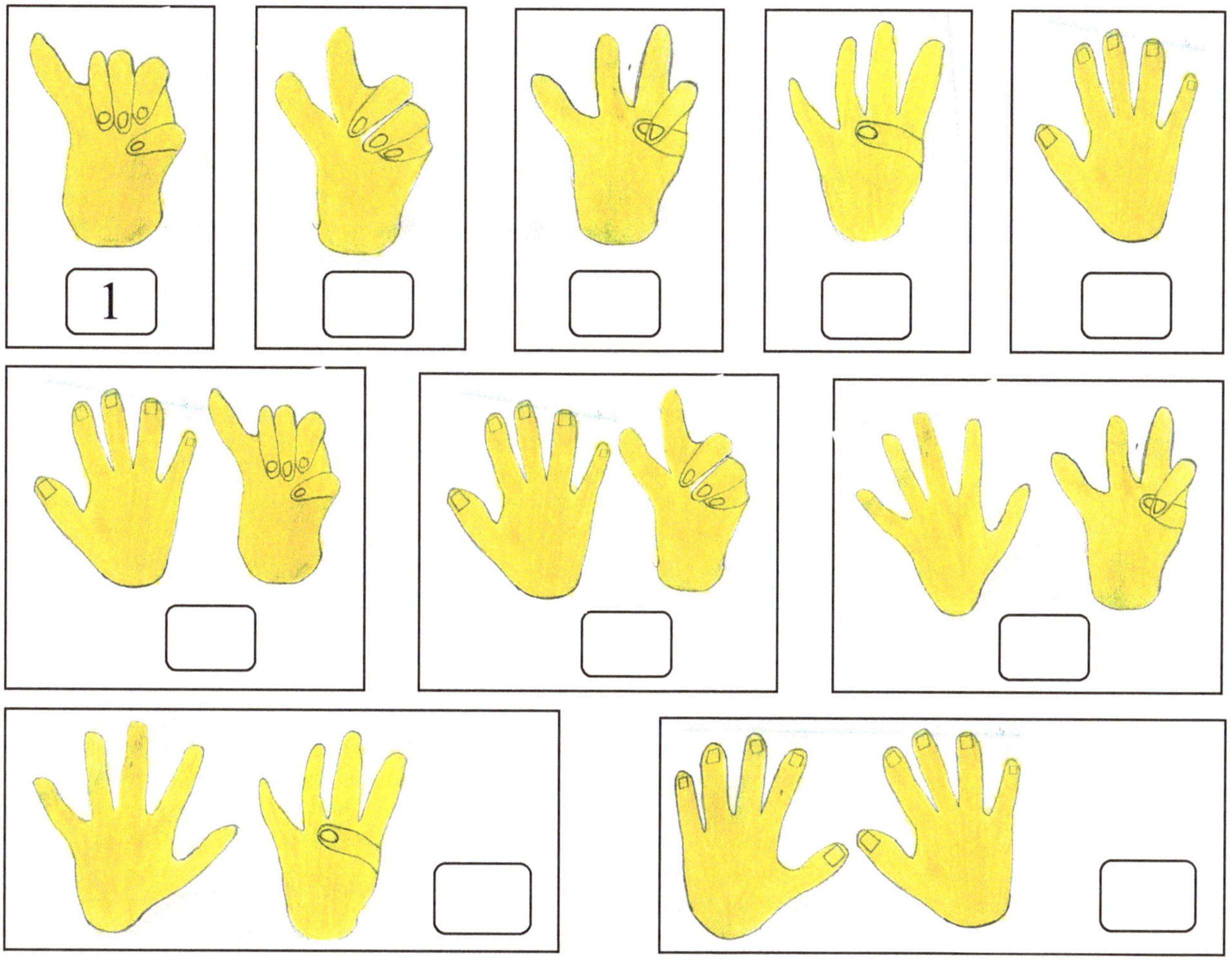

1

Count with fingers in an another way and fill in the missing places

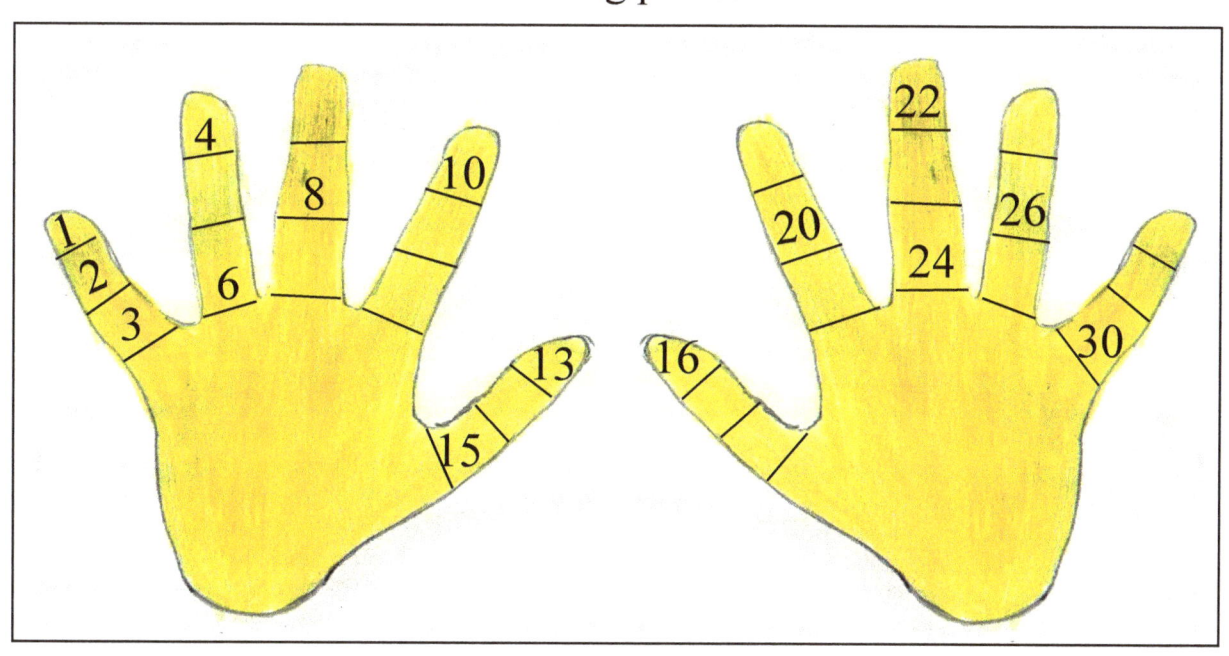

1	🐦	3	4	5	🍓	7	8	9	⭐
11	12	🚗	14	15	16	👧	18	🐀	20
🍎	22	23	24	🟩	26	🐌	28	29	🐺

Write the numbers that place for pictures

🐦	2	🍓		⭐		🚗		👧	
🐀		🍎		🟩		🐌		🐺	

Write these numbers with their correct words

25   22   26   23   21   27   29   30   28   24

Twenty one		Twenty two		Twenty three		Twenty four		Twenty five	
Twenty six		Twenty seven		Twenty eight		Twenty nine		Thirty	

Circle the next numbers in the brackets

23, 24, 25 (27, 26)	25, 26, 27 (28, 29)
26, 27, 28 (29, 30)	24, 25, 26 (29, 27)
21, 22, 23 (24, 26)	20, 21, 22 (23, 26)
22, 23, 24 (28, 25)	22, 24, 26 (28, 30)
27, 28, 29 (30, 26)	21, 23, 25 (29, 27)

Match and make digit words

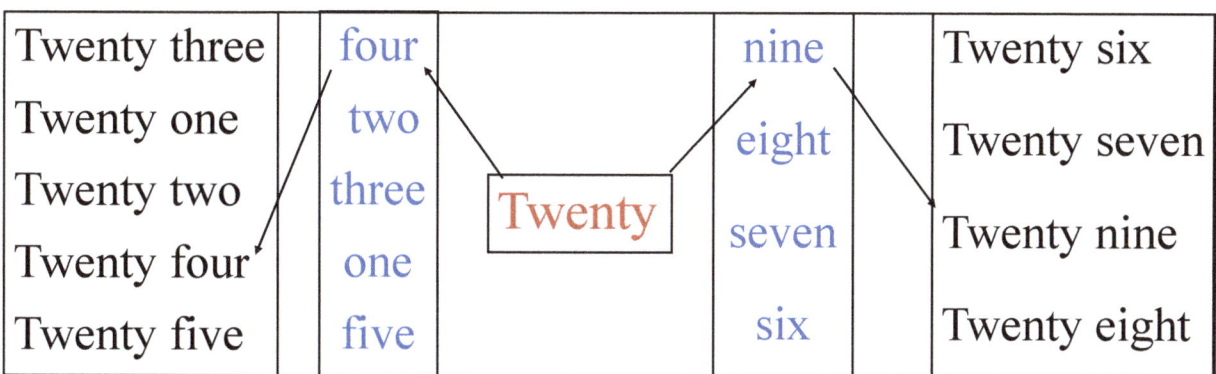

Match with the same group of sticks

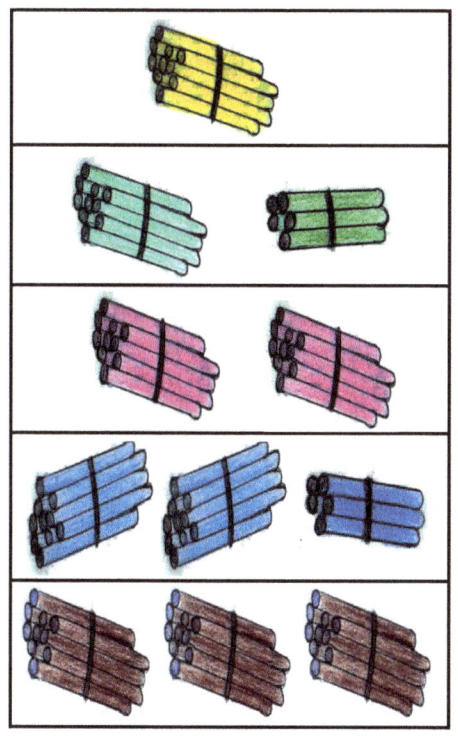
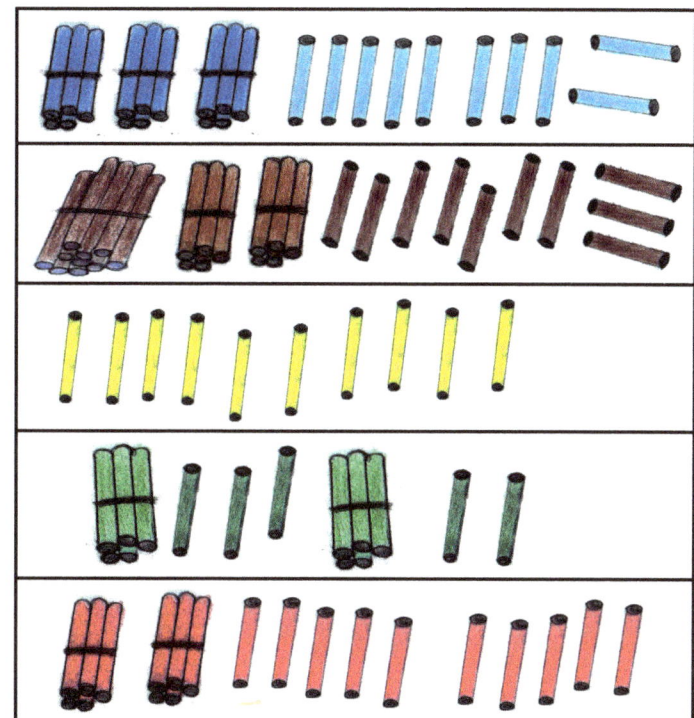

Look at the numbers and underline their correct words

21 Twenty two	Twenty one
22 Twenty two	Twenty five
23 Twenty one	Twenty three
24 Twenty five	Twenty four
25 Twenty five	Twenty two

26 Twenty four	Twenty six
27 Twenty two	Twenty seven
28 Twenty eight	Twenty two
29 Twenty nine	Twenty four
30 Twenty five	Thirty

Count beads with number frame and write numbers in the boxes

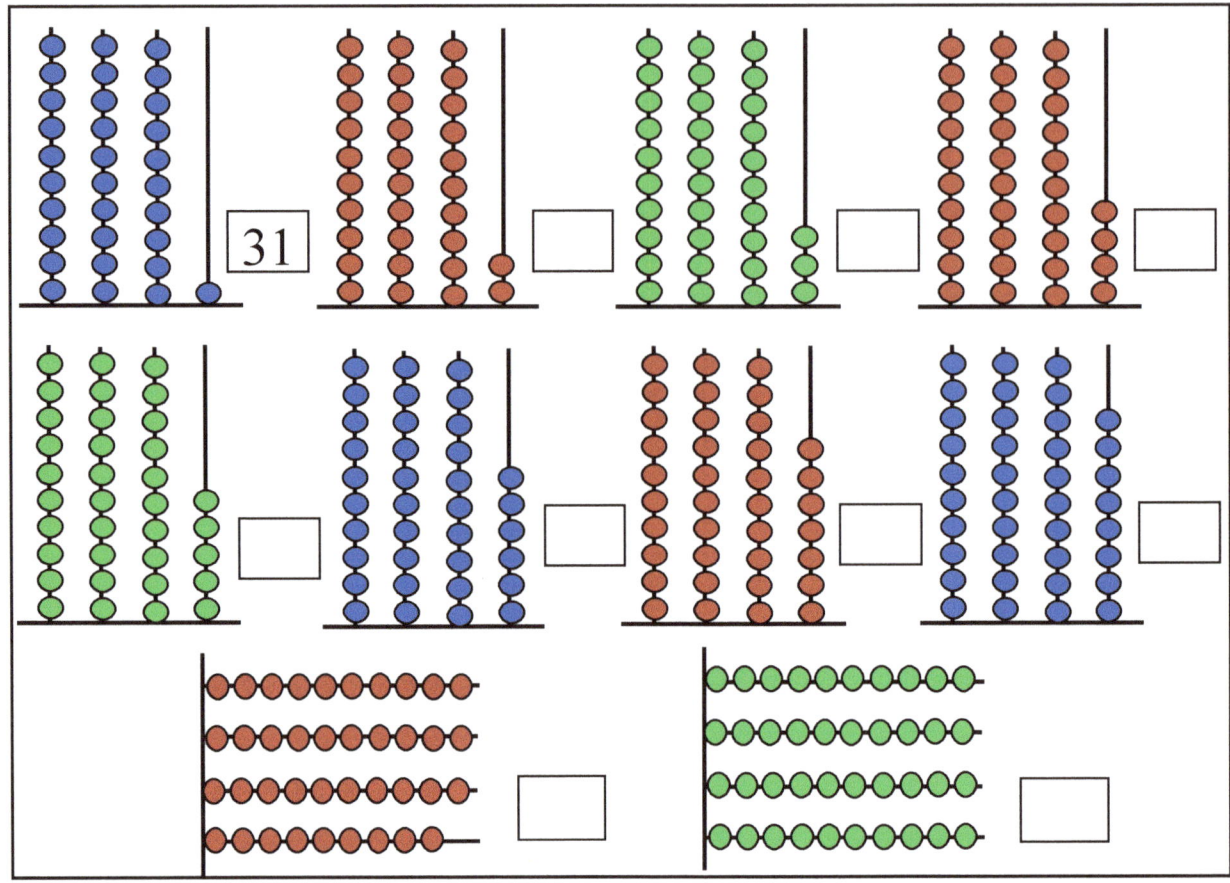

Choose the numbers in the brackets and fill in the missing spaces

32, 33, 34 (32, 34)	___, 37, ___ (38, 36)
___, 36, ___ (35, 37)	___, 34, ___ (33, 35)
___, 32, ___ (33, 31)	___, 38, ___ (37, 39)
___, 35, ___ (36, 34)	___, 39, ___ (38, 40)

Fill the numbers in the missing spaces from 31- 40 and 40 - 31

31		33			36		38	39	
	39		37	36		34	33		

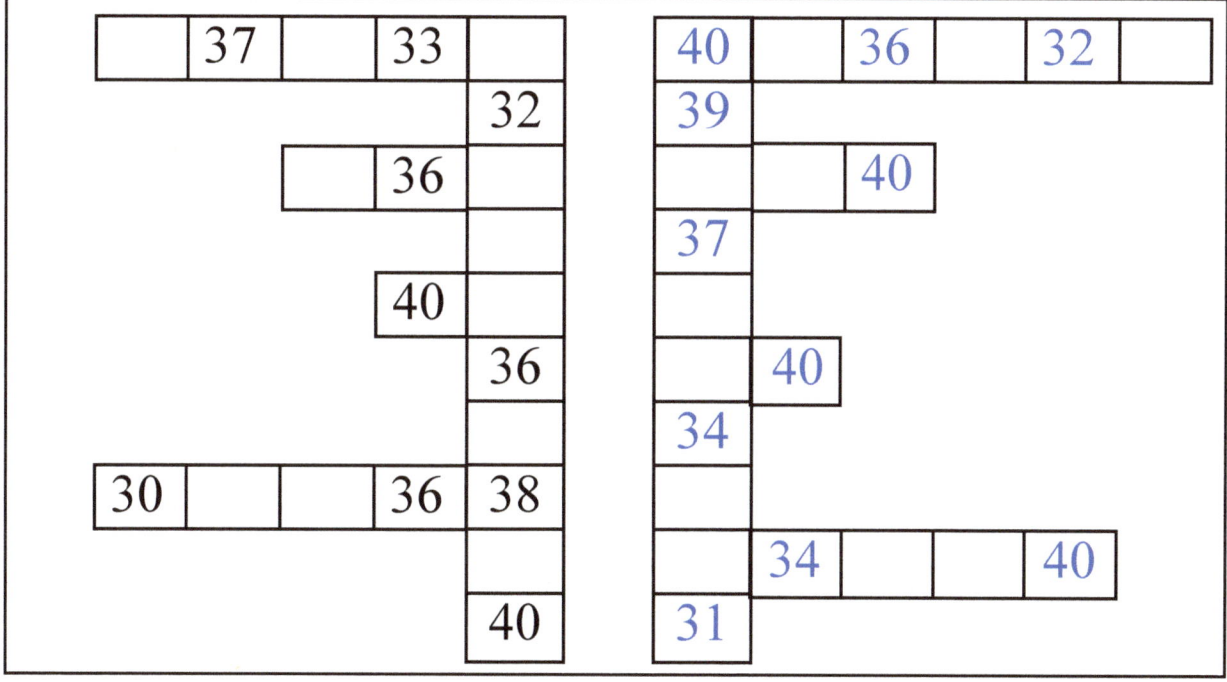

Match and make digit words 31–39

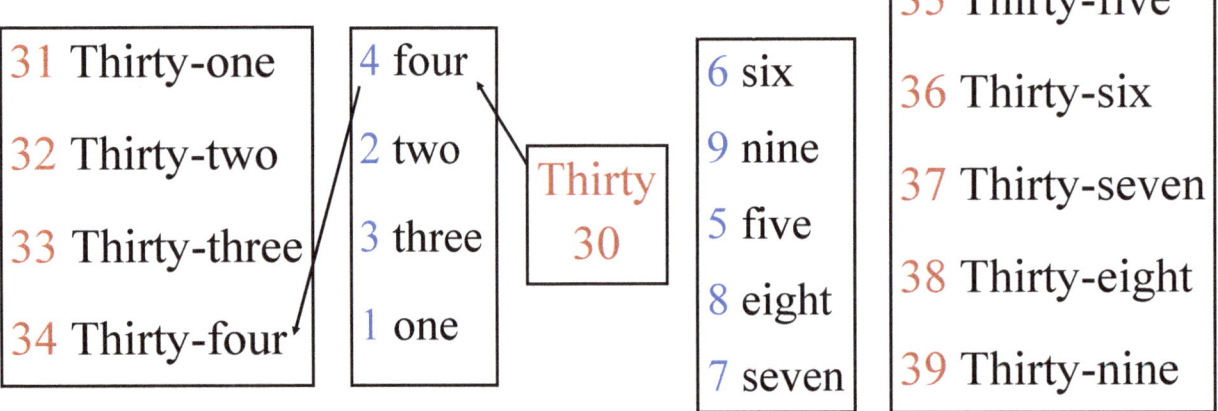

Select and colour the numbers from 31-40

31	32	13	54	33	54	34	65	35	60	42
16	36	72	37	38	67	39	70	40	90	67

Fill in the missing numbers and words

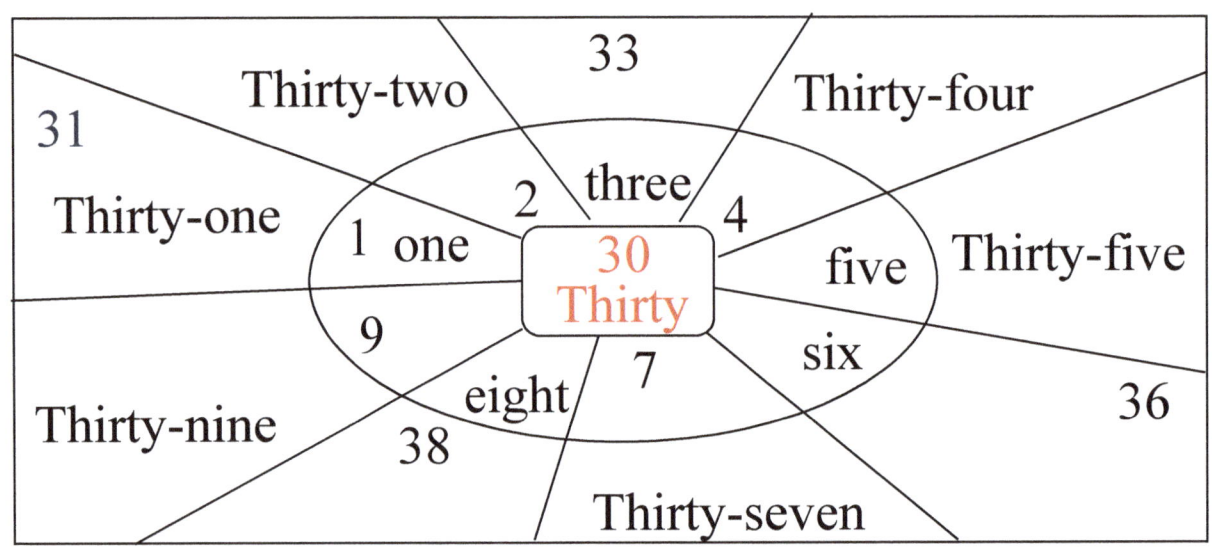

Create numbers 31 - 40 from the given numbers

531	312	313	434	553	356	373	483	239	420
31									

Find the same numbers from 41 - 50 and colour them

41	12	42	26
30	43	87	44
45	66	46	90
97	47	78	48
49	50	16	44

13	41	27	70	42	78	43	29
44	23	45	66	46	37	47	55
34	48	66	57	49	40	50	63

Find the pattern of the numbers and match with next number

41	42	43	44	45	46
38	40	42	44	46	48
50	48	46	44	42	40
30	33	36	39	42	45
49	47	45	43	41	39
15	20	25	30	35	40

38
37
47
50
45
48

Count the blocks and write the numbers in the boxes

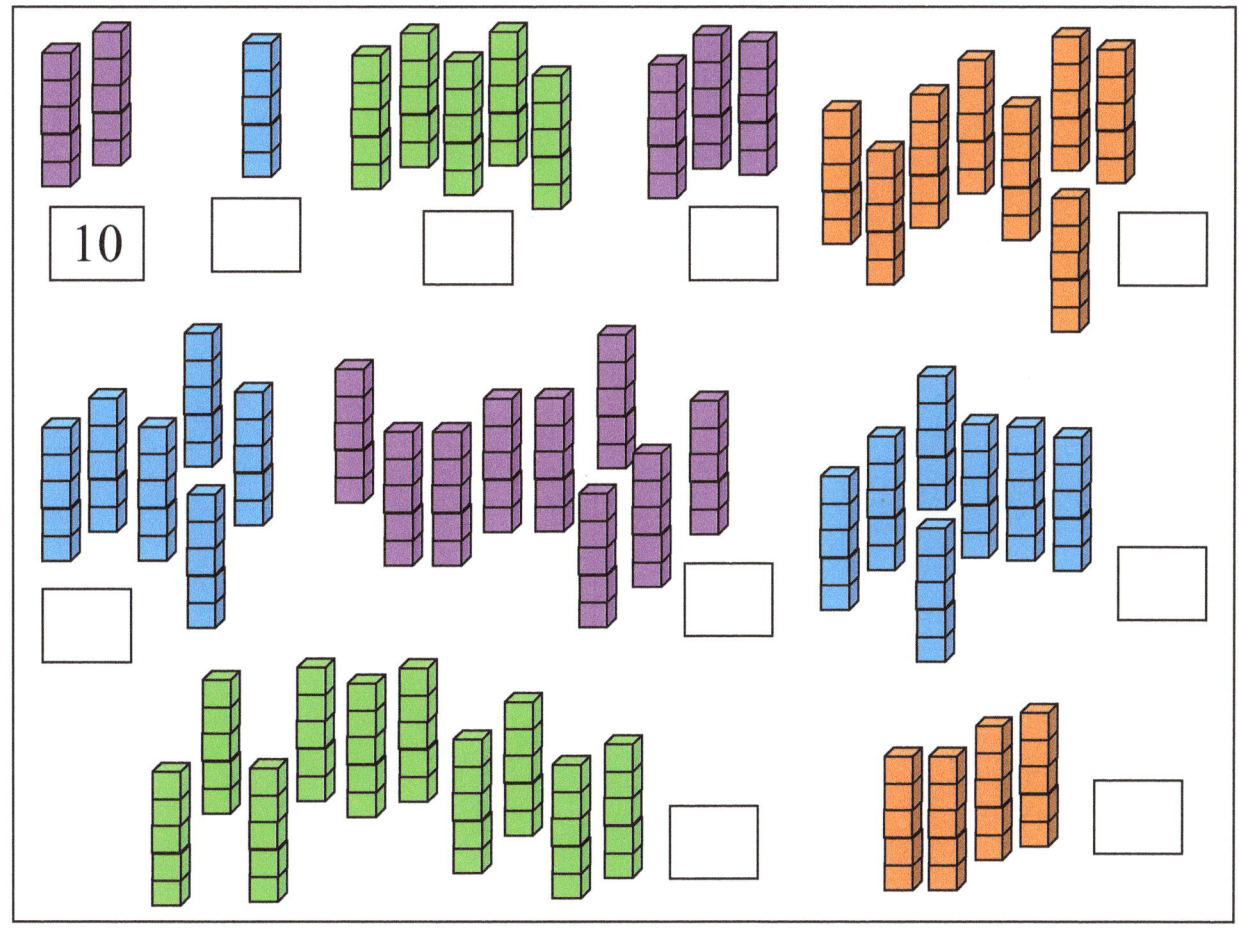

Look at the board and fill the numbers in the missing places

Match them (I have, Who has?)

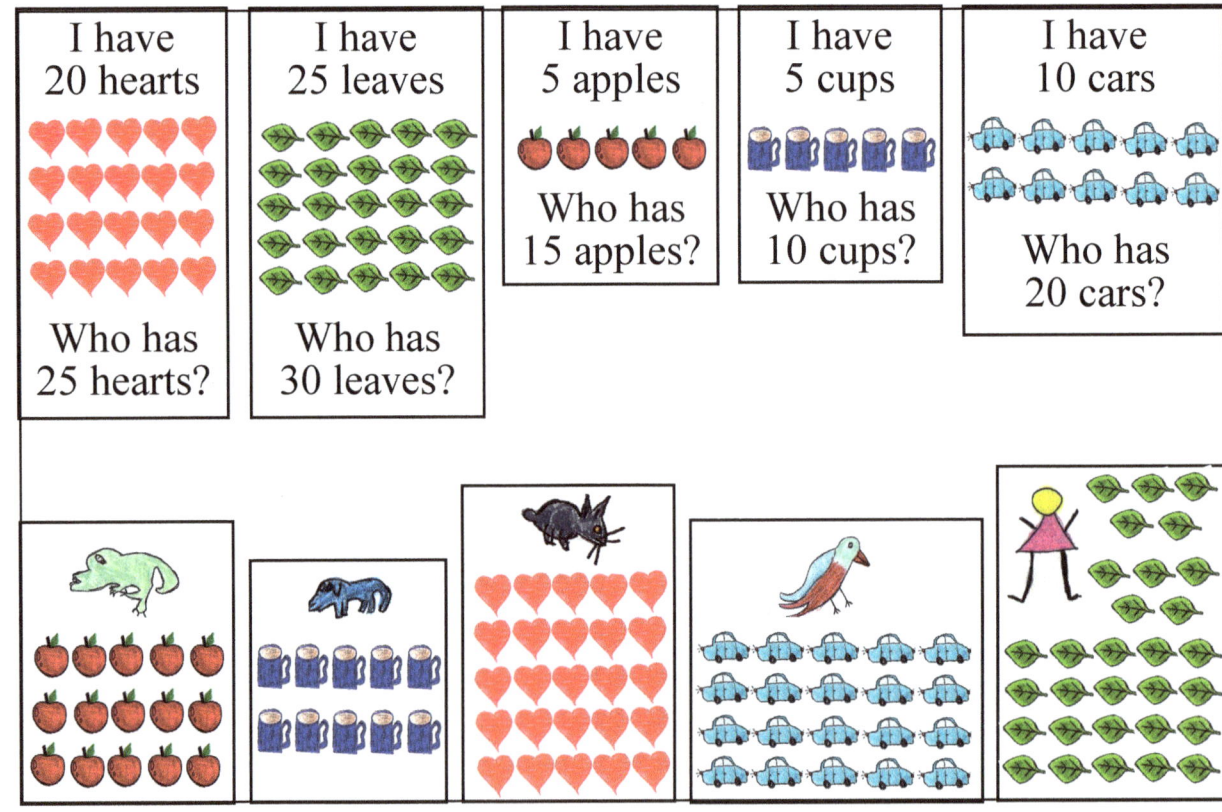

Count how many ones and tens and fill in the missing places

	Ones		Tens
(10 dots)	10	(1 stack)	1
(20 dots)		(2 stacks)	
(30 dots)		(3 stacks)	
(40 dots)		(4 stacks)	
(50 dots)		(5 stacks)	

Ones	Tens
	1
20	
	3
40	

Tens	Ones
2	
	30
4	
	50

Fill in the missing spaces and make dight words

41	F	o	r	t	y	-	o		n	e	
42	F	o	r	t		-	t	w			
43	F	o	r			-	t	h	r		e
44	F	o				-	f		u	r	
45	F					-	f	i		e	

46	F	o	r	t	y	-		i	x		
47		o	r	t	y	-	s		v		n
48			r	t	y	-		i	g	h	t
49				t	y	-	n		n	e	
50	F			f		y					

Make numbers chart 1 - 50
Fill in the missing numbers and words

	11	Twenty one	Thirty one	Forty one
One				
2		Twenty two	32	42
	Twelve			
		23	Thirty three	Forty three
Three	Thirteen			
4		Twenty four	34	Forty four
	Fourteen			
	15	Twenty five	Thirty five	45
Five				
Six	Sixteen	Twenty six	Thirty six	46
		27	Thirty seven	Forty seven
Seven	Seventeen			
8	18	Twenty eight	38	Forty eight
		29	Thirty nine	Forty nine
Nine	Nineteen			
10	Twenty	30	Forty	50

What number I am (Circle the answers)

1) I am 2 more than 12 (15, 14)

2) I am 3 more than 24 (27, 29)

3) I am 5 more than 35 (40, 45)

4) I am 4 more than 46 (50, 54)

5) I am 10 more than 30 (20, 40)

6) I am 5 more than 25 (35, 30)

1) I am 2 less than 18 (20, 16)

2) I am 3 less than 43 (47, 40)

3) I am 5 less than 50 (45, 40)

4) I am 10 less than 40 (30, 50)

Fill the numbers by pattern

Skip by 2 and colour the numbers (Top to down)

46	48	50	52
47	49	51	53
48	50	52	54
49	51	53	55
50	52	54	56
51	53	55	57
52	54	56	58
53	55	57	59
54	56	58	60

Sort and write the numbers

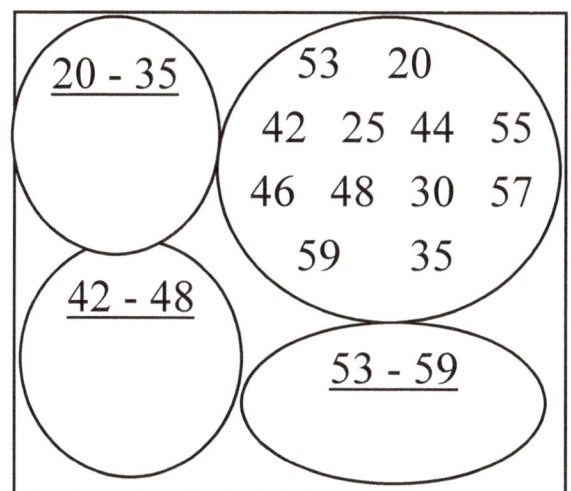

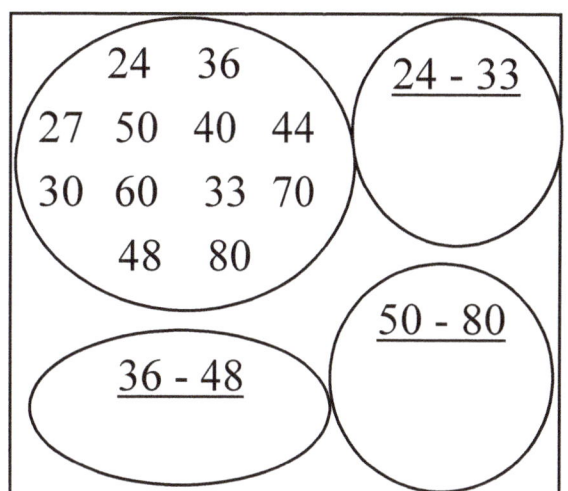

Circle the number which is different from the pattern

51,	52,	53,	59,	54
52,	54,	56,	58,	79
58,	57,	56,	85,	55
77,	59,	57,	55,	53

53,	44,	55,	57,	59
45,	50,	55,	60,	90
58,	70,	56,	54,	52
51,	81,	54,	57,	60

Find the given numbers and circle around them

1) 53, 54, 55, 56, 57
2) 55, 56, 57, 58, 59
3) 56, 57, 58, 59, 60
4) 57, 56, 55, 54, 53
5) 60, 59, 58, 57, 56

59	60	59	58	57	56
60	53	55	56	52	60
59	54	51	53	57	55
58	55	56	57	58	59
57	56	56	51	54	53
56	57	56	55	54	53

59	57	55	53	51	52
59	57	55	57	59	54
53	60	51	57	55	56
51	53	55	57	59	58
57	56	55	54	53	60
60	58	56	54	52	60

1) 51, 53, 55, 57, 59
2) 59, 57, 55, 57, 59
3) 52, 54, 56, 58, 60
4) 60, 58, 56, 54, 52
5) 57, 56, 55, 54, 53

Make a group with numbers and their words

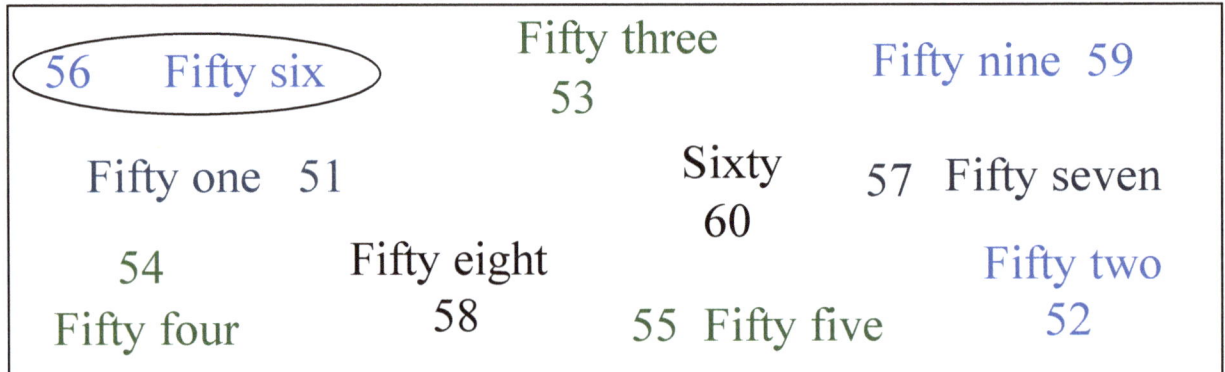

Connect numbers and their words in order 51 - 60

51 Fifty-one   54 Fifty-four   57 Fifty-seven   56 Fifty-six

53 Fifty-three   52 Fifty-two   55 Fifty-five

60 Sixty   59 Fifty nine   58 Fifty eight

Match them (I have, Who has?)

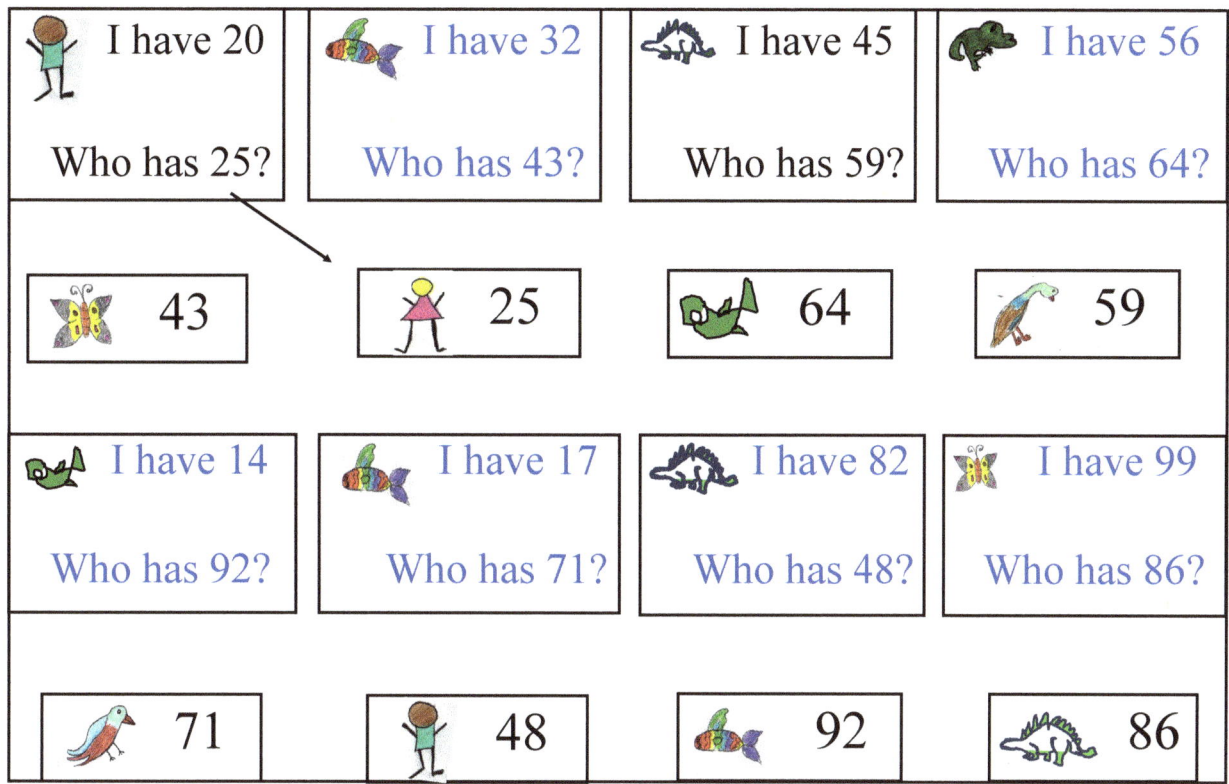

(Circle the answers)

1) I am not 58 (58, 68)	1) I am twenty five (25, 45)
2) I am not 54 (54, 44)	2) I am thirty nine (45, 39)
3) I am not 33 (43, 33)	3) I am fifty five (53, 55)
4) I am not 52 (27, 52)	4) I am fifty four (54, 59)
5) I am not 49 (49, 91)	5) I am nineteen (19, 90)
6) I am not 36 (36, 39)	6) I am fifty nine (95, 59)
7) I am not 55 (66, 55)	7) I am sixty (60, 75)
8) I am not 60 (60, 66)	8) I am fifty eight (58, 46)

Find the pattern of the numbers and fill in the missing places

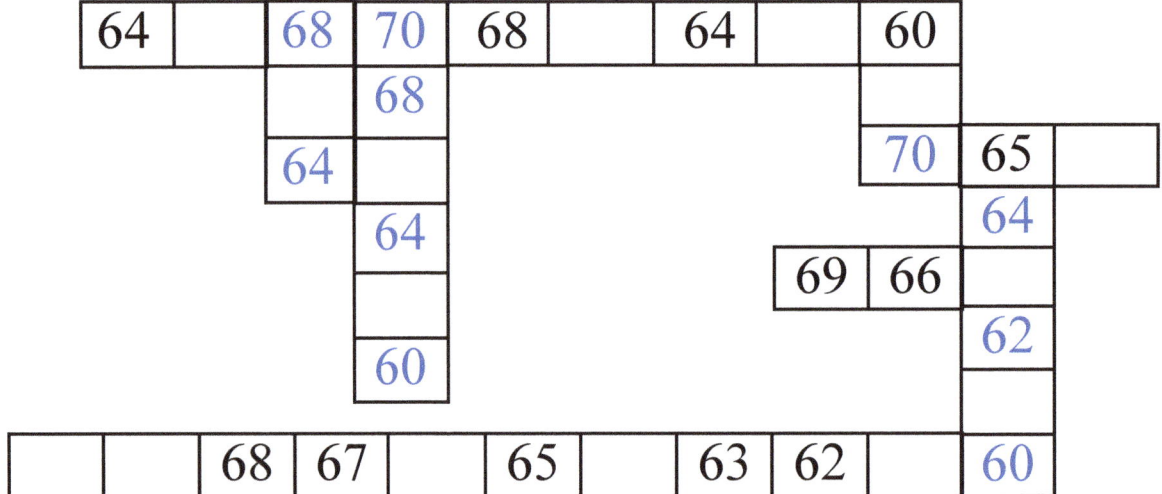

Connect next numbers with pattern

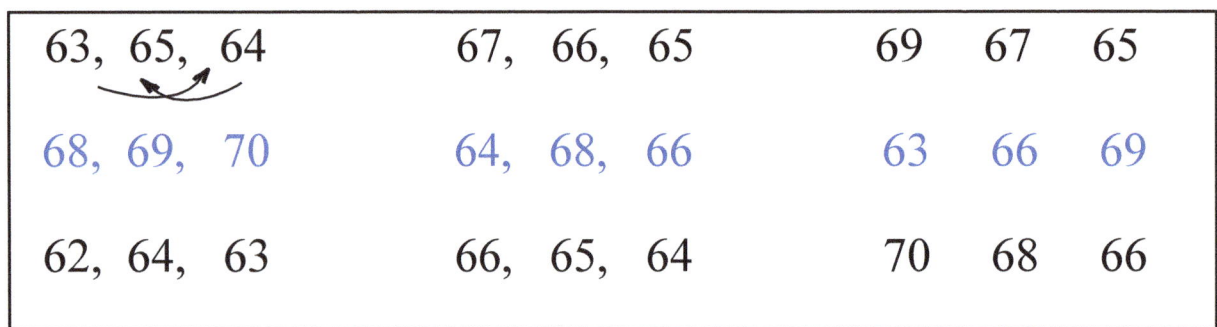

Arrange these numbers in order from 61 – 70

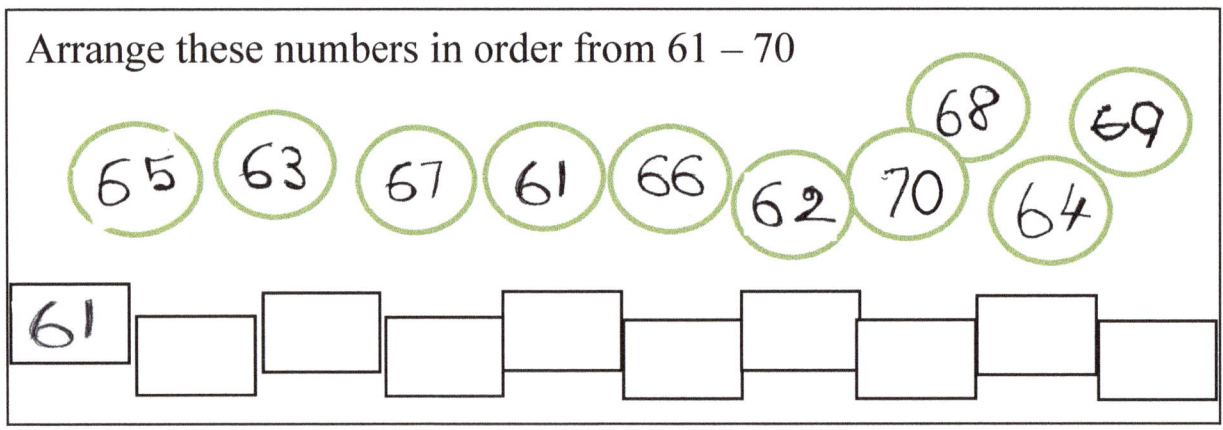

What is the number in between?

| 63 ___ 65 | 61 ___ 63 | 67 ___ 69 | 68 ___ 70 |
| 66 ___ 68 | 65 ___ 67 | 62 ___ 64 | 64 ___ 66 |

Climb and descend form the ladders

Connect the numbers in the order

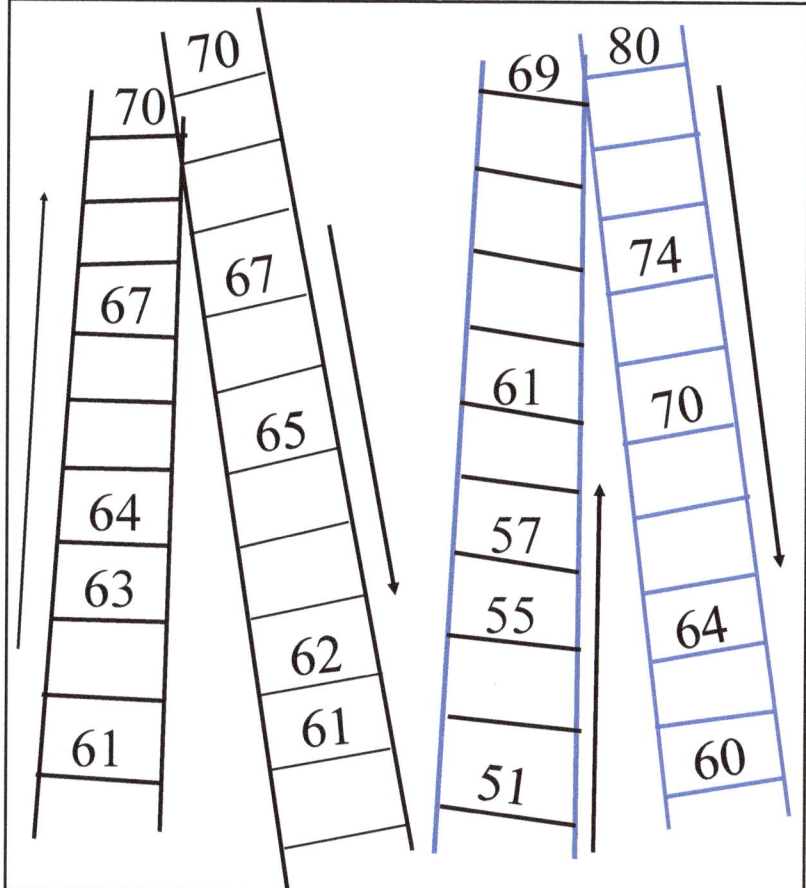

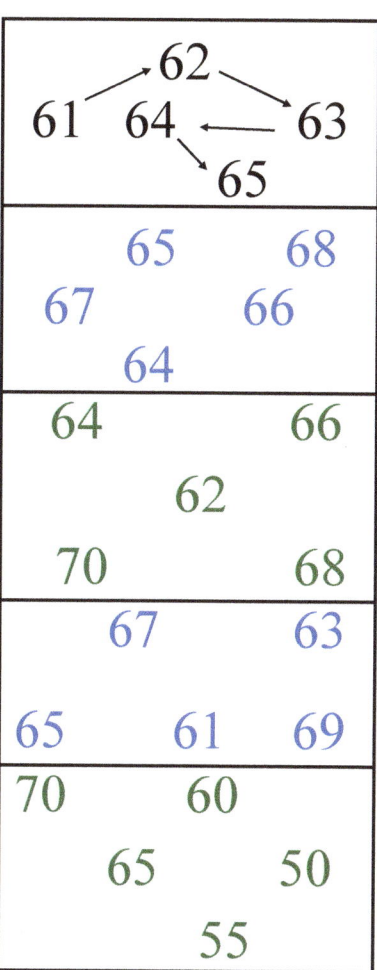

Arrange these numbers from small to big

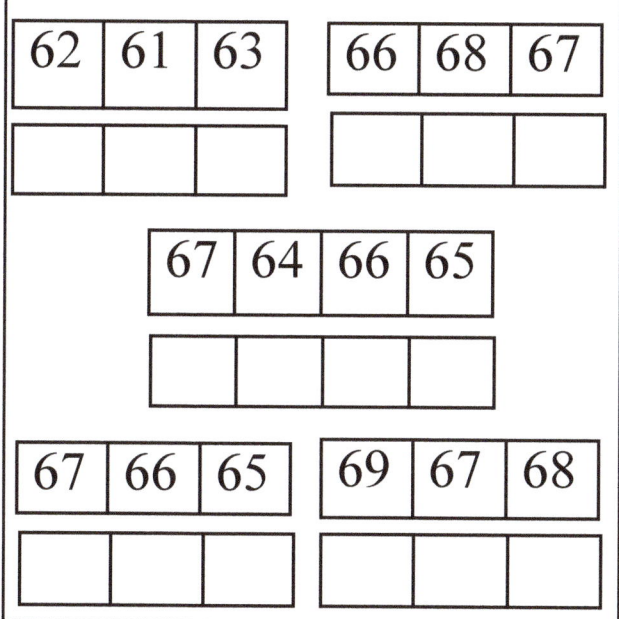

Arrange these numbers from big to small

Arrange these letters and make digit words

t	i	s	x	y		w	o	t
s	i	x	t	y	-	t	w	o

y	i	s	x	t		n	o	e

t	x	s	i	y		v	e	f	i

t	x	y	i	s		f	o	r	u

t	i	s	x	y		s	x	i

y	x	s	i	t		e	h	t	r	e

s	t	i	y	x		s	n	e	v	e

t	i	s	x	y		i	n	e	n

y	i	t	x	s		e	i	t	h	g

t	e	s	v	y	n	e

Pick up the numbers and fill the bingo chart

60 15 65 69 30
20 63 54 69 50

B	I	N	G	O
63		57		10
	56	60	20	
67	58		25	30
		66		40
71	62		35	

Find the pattern and fill in the missing spaces

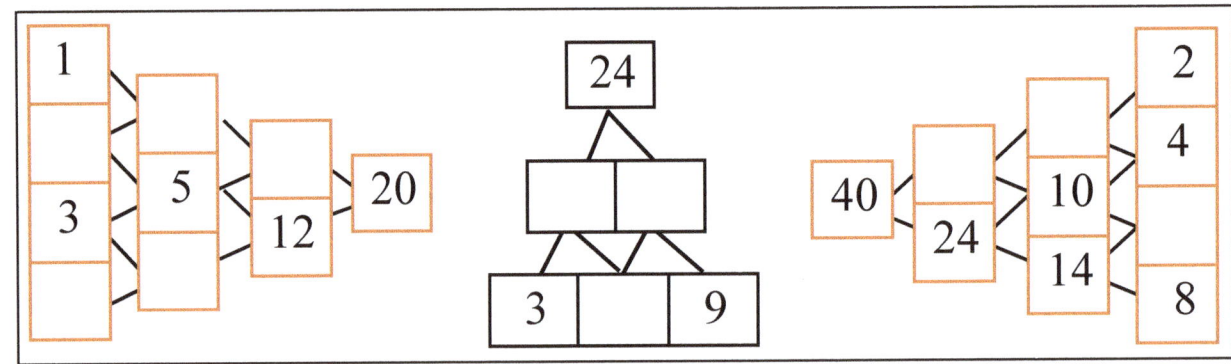

Read numbers and colour the beads

Match the same set of the numbers

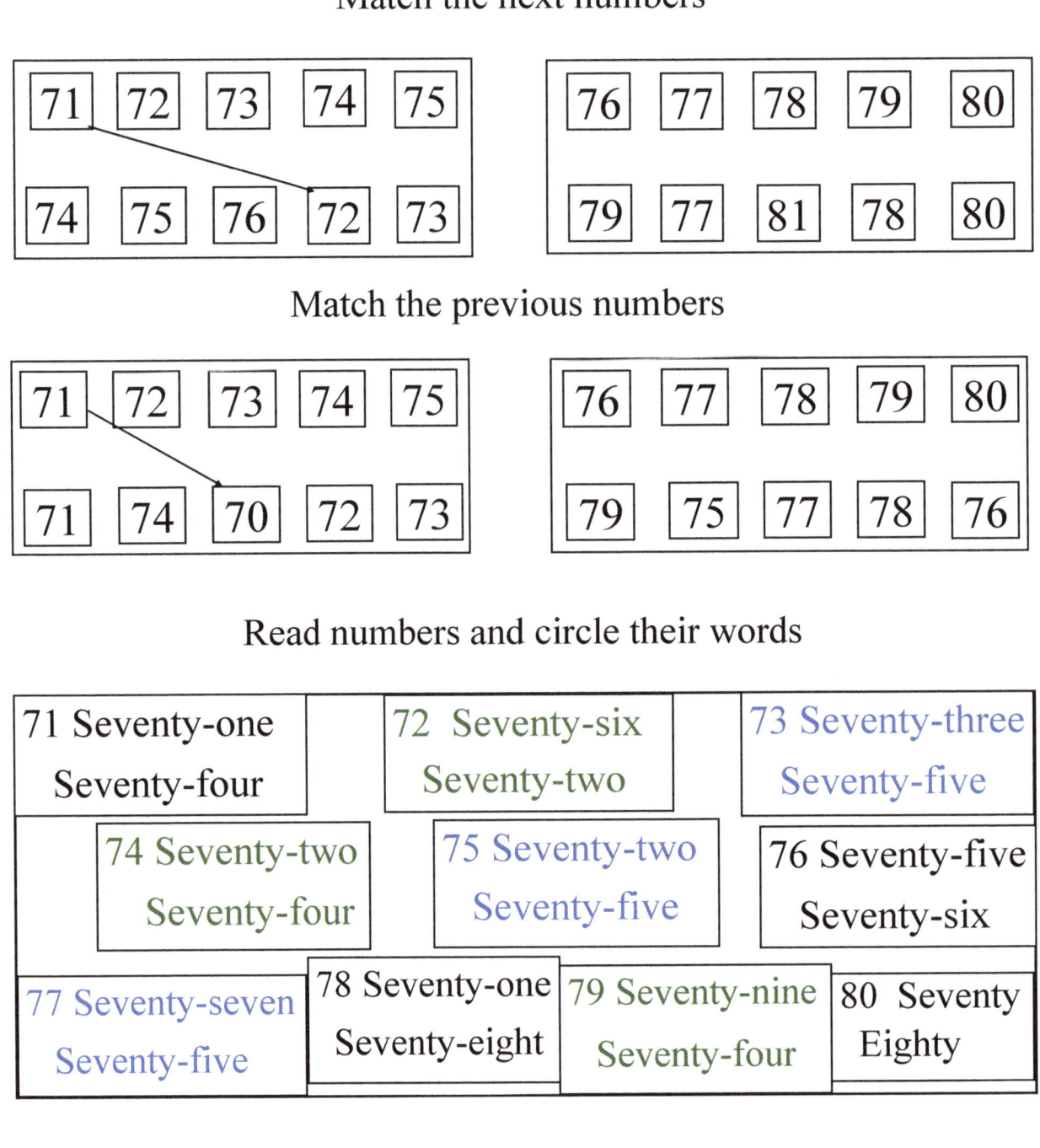

Find the next numbers in order

83, 84, ___	82, 83, ___	83, 82, ___
85, 86, ___	87, 88, ___	87, 86, ___
84, 85, ___	88, 89, ___	84, 83, ___

Find the numbers in between

83, ___, 85	84, ___, 86	85, ___, 87
86, ___, 88	87, ___, 89	80, ___, 82
82, ___, 84	81, ___, 83	88, ___, 90

Colour the numbers 81 - 90 and 90 - 81

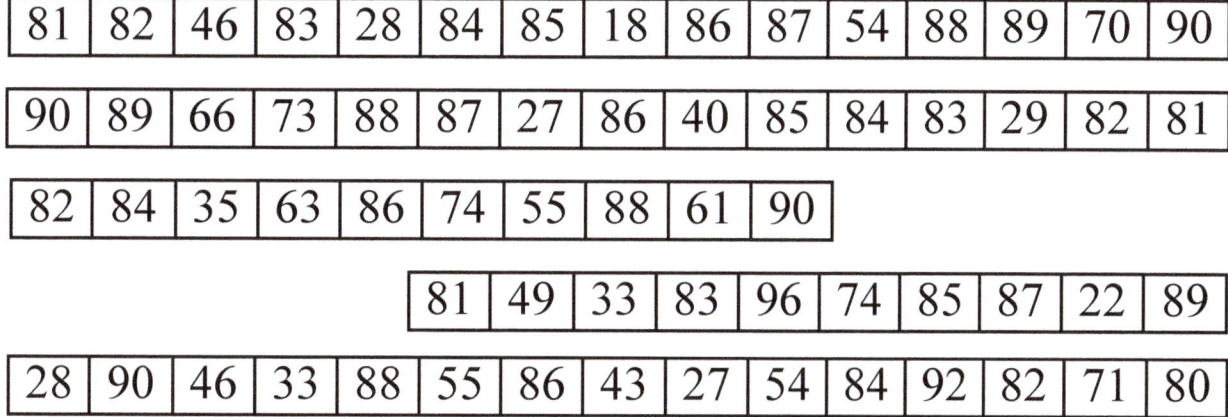

Find the previous numbers

___, 85, 86	___, 84, 85	___, 89, 88
___, 82, 83	___, 81, 82	___, 82, 81
___, 87, 88	___, 83, 84	___, 87, 86

Make two digit numbers from given numbers

8 6 3	3 1 9	6 1 8	9 2 7
86			
83			
63			
68			
36			
38			

Find the previous and next numbers

___ , 87, ___

___ , 82, ___

___ , 85, ___

___ , 84, ___

___ , 89, ___

___ , 83, ___

___ , 86, ___

___ , 88, ___

Fill the missing letters and make digit words

81	E	i	g	h	t	y	-	o	n		
82	E	i	g	h	t		-		w	o	
83	E	i	g	h			-	t	h	e	e
84	E	i	g				-	o	u		
85	E	i					-	f		e	

86					t	y	-	s			
87				h	t	y	-		e	e	n
88			g	h	t	y	-	e		g	
89		i	g	h	t	y	-	n		e	

| 90 | N | i | | e | t |

Trace the numbers 81 - 90 and 90 - 81

81	82
83	84
85	86
87	88
89	90

90	89	88	87	86
85	84	83	82	81

Read the digit words and circle their numbers

Ninety two	93,	92
Ninety four	94,	92
Ninety five	93,	95
Ninety one	91,	93
Ninety seven	97,	99

Ninety three	93,	97
Ninety six	94,	96
Ninety nine	93,	99
One hundred	100,	70
Ninety eight	98,	68

Find the numbers value and write them in the correct places

32

Tens	Ones
3	2

66

Tens	Ones

43

Tens	Ones

55

Tens	Ones

25

Tens	Ones

84

Tens	Ones

Hundred	Tens	Ones
		1
	1	0
1	0	0

Hundred	Tens	Ones	Numbers
	2	8	2 8
1	0	0	
	7	8	
	9	9	
	6	4	

100

Hundred	Tens	Ones

72

Tens	Ones

90

Tens	Ones

93

Tens	Ones

Numbers	Hundred	Tens	Ones
25			
9			
30			
88			
100			
91			

Look at the red coloured numbers from the first numbers chart and colour them

Connect the numbers in order from 2 - 50

1	11	21	31	41
2	12	22	32	42
3	13	23	33	43
4	14	24	34	44
5	15	25	35	45
6	16	26	36	46
7	17	27	37	47
8	18	28	38	48
9	19	29	39	49
10	20	30	40	50

1	11	21	31	41
2	12	22	32	42
3	13	23	33	43
4	14	24	34	44
5	15	25	35	45
6	16	26	36	46
7	17	27	37	47
8	18	28	38	48
9	19	29	39	49
10	20	30	40	50

[2] → [4]
[6]   [8]
[12] [10]
[16] [14]
[18] [20]

Circle the red coloured numbers

51	61	71	81	91
52	62	72	82	92
53	63	73	83	93
54	64	74	84	94
55	65	75	85	95
56	66	76	86	96
57	67	77	87	97
58	68	78	88	98
59	69	79	89	99
60	70	80	90	100

[22] [30] [32] [38] [46] [48]
[24]   [26] [36] [42] [44]
   [28]   [34]   [40]   [50]

Follow the arrow and fill the numbers with pattern

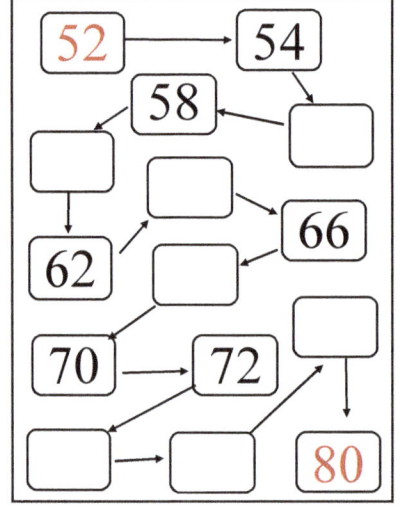

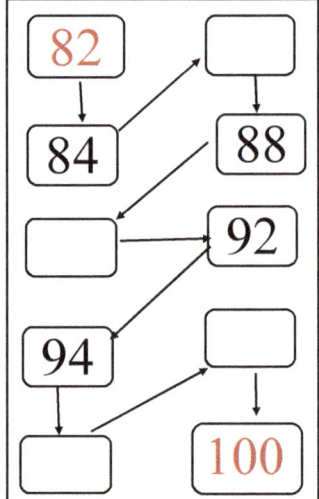

Fill in the missing numbers from 51- 70

51			54			57			60
		63		65			68	69	

Circle the red coloured numbers

1	2	3	4	5	6	7	8	9	10
11	12	13	14	15	16	17	18	19	20
21	22	23	24	25	26	27	28	29	30
31	32	33	34	35	36	37	38	39	40
41	42	43	44	45	46	47	48	49	50

51	52	53	54	55	56	57	58	59	60
61	62	63	64	65	66	67	68	69	70
71	72	73	74	75	76	77	78	79	80
81	82	83	84	85	86	87	88	89	90
91	92	93	94	95	96	97	98	99	100

Connect the numbers from 1 - 99 with pattern

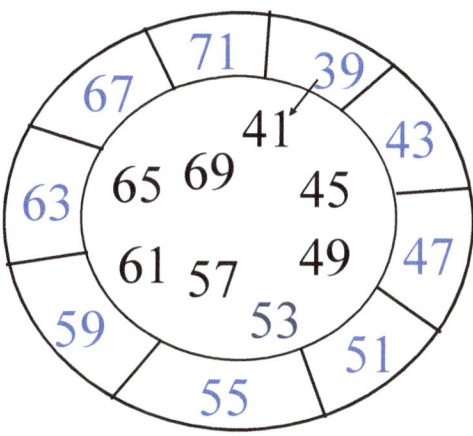

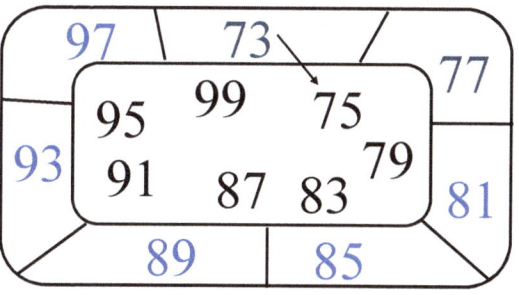

Find the digit words and circle them

t	w	e	n	t	y	
h		i	i		t	y
i		g	n		x	t
r		h	e		i	r
t		t	t		s	o
y		y	y			f
	y	t	f	i	f	
s	e	v	e	n	t	y

Sort and write the numbers from 91 - 100

32, 91, 27
76, 43, 92
54, 93, 39
15, 94, 19
95, 64, 27
66, 96, 69
97, 53, 39
25, 19, 98
29, 79, 99
10, 100, 90

91

Circle the red coloured numbers

1	2	3	4	5	6	7	8	9	10
11	12	13	14	15	16	17	18	19	20
21	22	23	24	25	26	27	28	29	30
31	32	33	34	35	36	37	38	39	40
41	42	43	44	45	46	47	48	49	50
51	52	53	54	55	56	57	58	59	60
61	62	63	64	65	66	67	68	69	70
71	72	73	74	75	76	77	78	79	80
81	82	83	84	85	86	87	88	89	90
91	92	93	94	95	96	97	98	99	100

Fill in the missing spaces

5	10
	20
25	
	40
55	
	70
75	
	100

Count these sticks and write the numbers

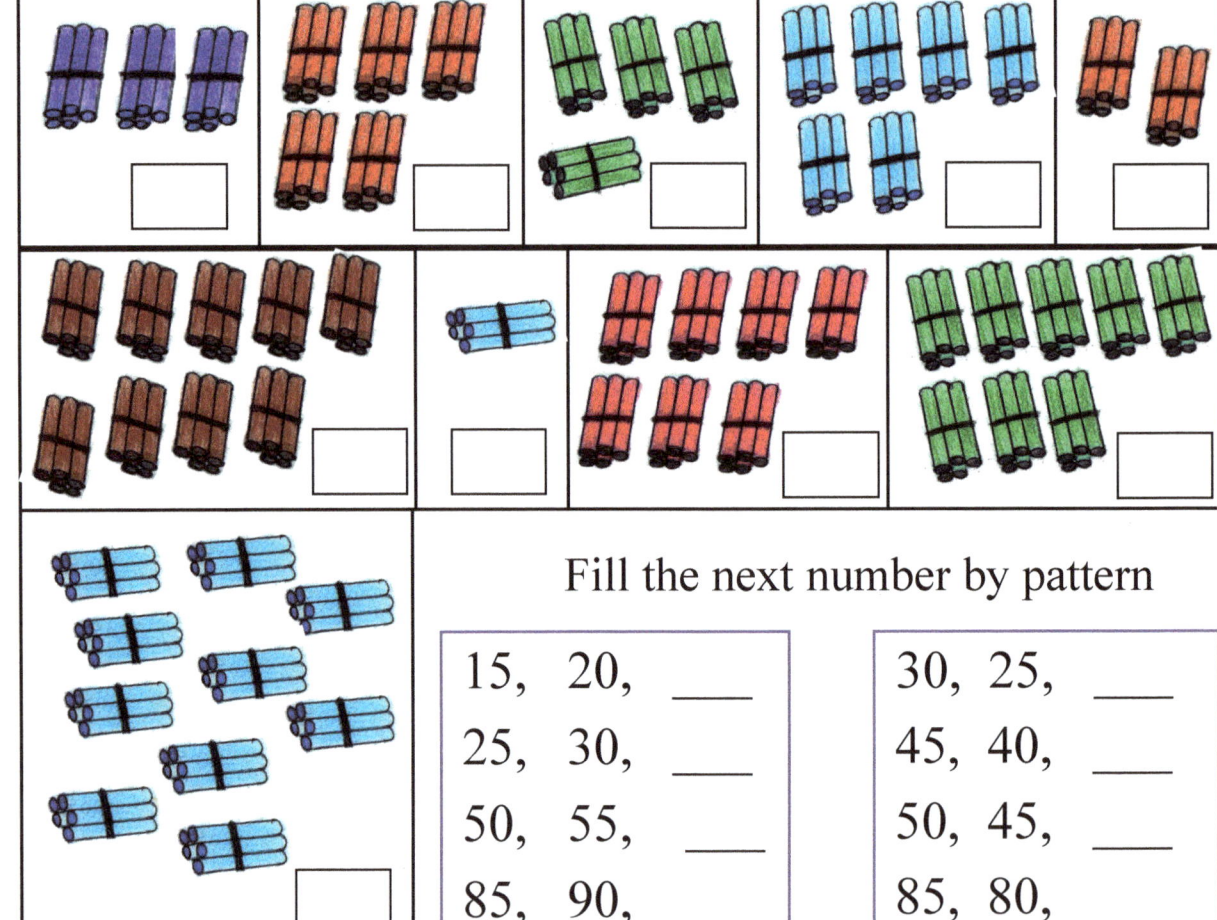

Fill the next number by pattern

15, 20, ___
25, 30, ___
50, 55, ___
85, 90, ___

30, 25, ___
45, 40, ___
50, 45, ___
85, 80, ___

Count with ten groups and write the numbers in the boxes

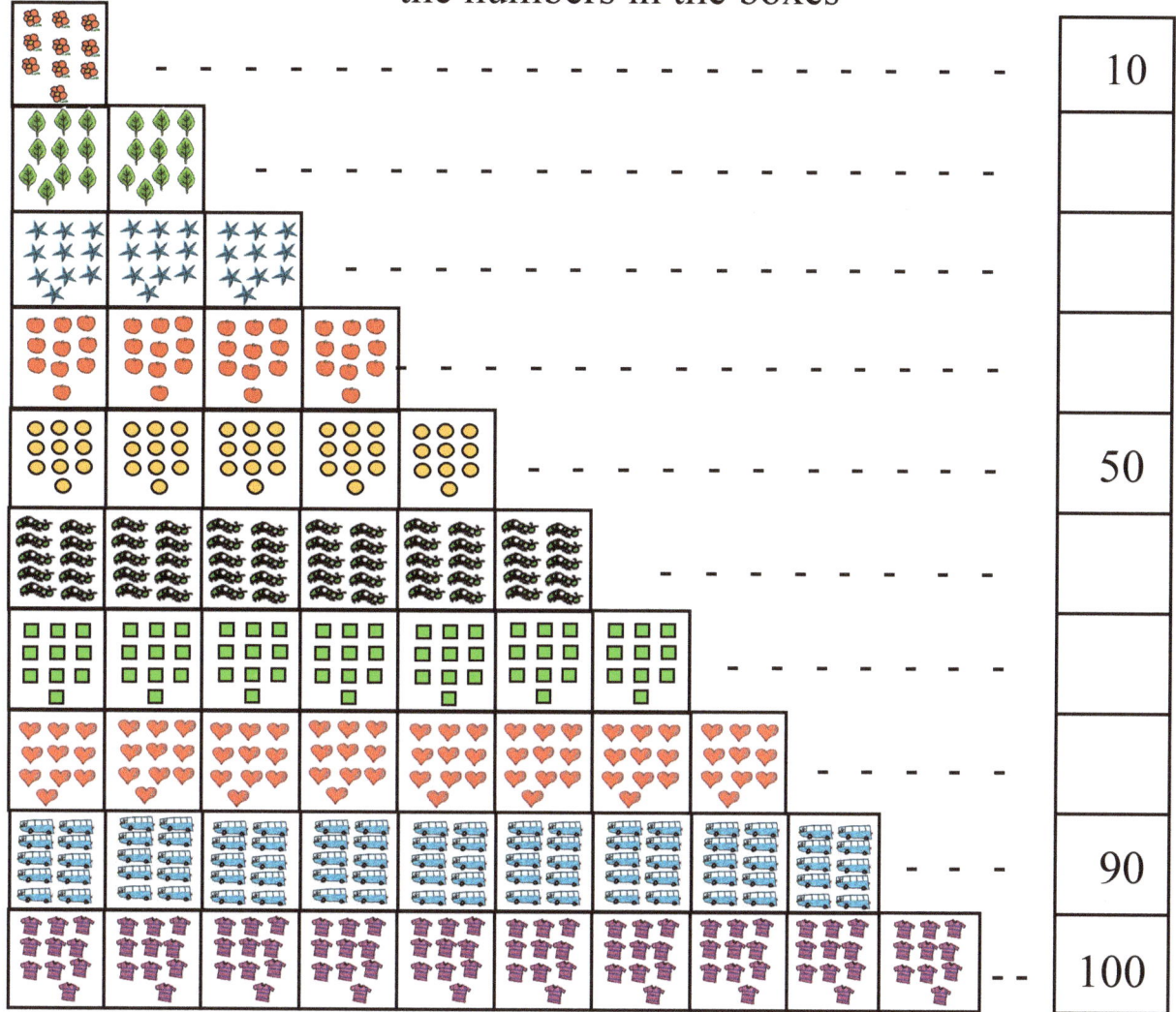

Circle the red coloured numbers

1	2	3	4	5	6	7	8	9	⑩
11	12	13	14	15	16	17	18	19	20
21	22	23	24	25	26	27	28	29	30
31	32	33	34	35	36	37	38	39	40
41	42	43	44	45	46	47	48	49	50
51	52	53	54	55	56	57	58	59	60
61	62	63	64	65	66	67	68	69	70
71	72	73	74	75	76	77	78	79	80
81	82	83	84	85	86	87	88	89	90
91	92	93	94	95	96	97	98	99	100

Skip by 10 and colour them

95	90	85	80	75
90	85	80	75	70
85	80	75	70	65
80	75	70	65	60
75	70	65	60	55
70	65	60	55	50
65	60	55	50	45
60	55	50	45	40

Write and count how many tens

100	10		10		10		10		10	10
90		10		10		10		10	- -	
80		10			10		- - - -			
70	10			10			10	- - - - - - -		
60			10		10	- - - - - - - - - -				
50		10		10	- - - - - - - - - - - -					5
40	10			10	- - - - - - - - - - - - - -				
30		10	- - - - - - - - - - - - - - - - - -						
20		10	- - - - - - - - - - - - - - - - - - - -						
10		- -							1

Fill in the missing places and make a number chart 51-100

51 Fifty-one	Sixty-one	71	Eighty-one	91
Fifty-two	62	Seventy-two	82	Ninety-two
53	Sixty-three	73	Eighty-three	Ninety-three
54	Sixty-four	Seventy-four	Eighty-four	94
Fifty-five	Sixty-five	Seventy-five	85	Ninety-five
Fifty-six	66	Seventy-six	86	Ninety-six
57	Sixty-seven	77	87	Ninety-seven
Fifty-eight	68	78	Eighty-eight	98
Fifty-nine	Sixty-nine	Seventy-nine	Eighty-nine	99
Sixty	Seventy	Eighty	Ninety	One hundred

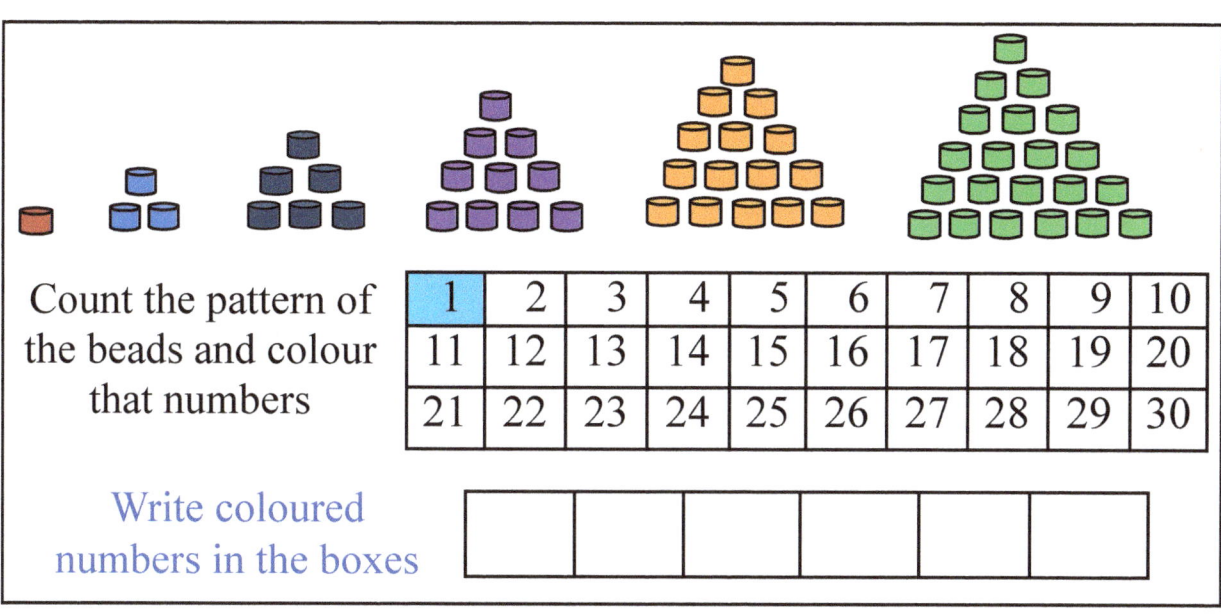

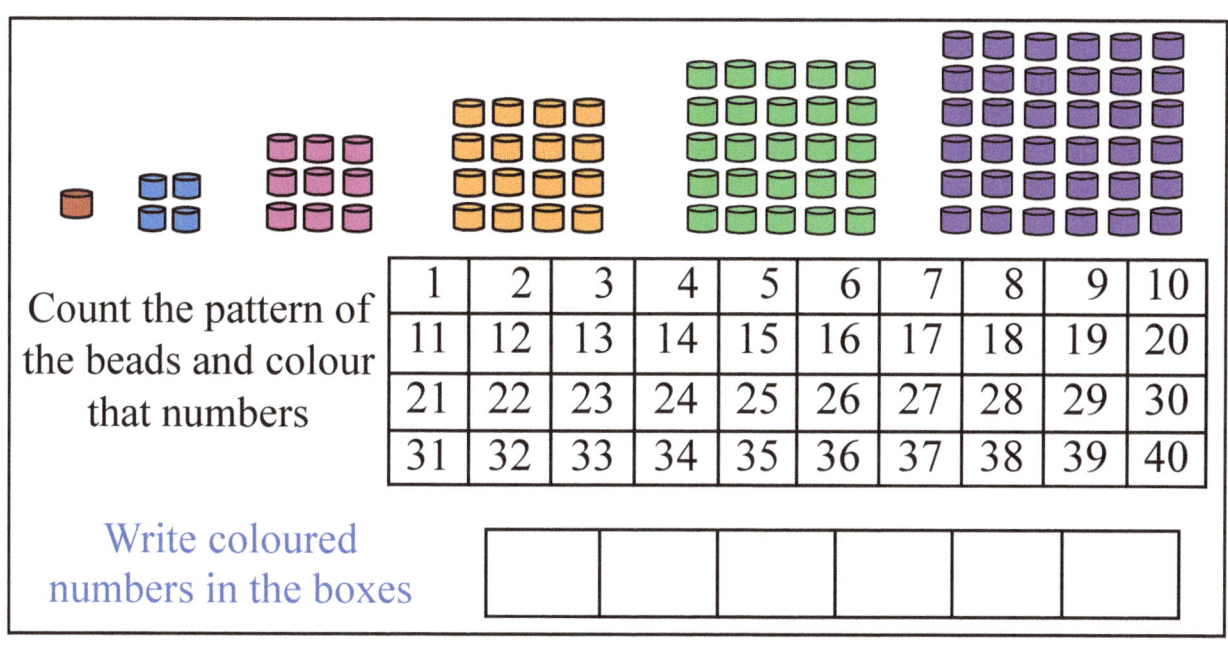

Circle the correct answers

How many numbers in this clock face?   12   24

How many one digit numbers in this clock face?   10   9

How many numbers in this clock face?   12   24

How many two digit numbers in this clock face?   10   15

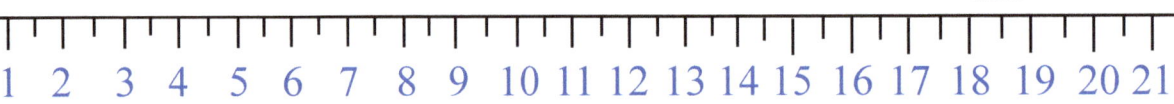

How many numbers in this ruler?   21   25

How many one digit numbers in this ruler?   6   9

January   2021

Monday	Tuesday	Wednesday	Thursday	Friday	Saturday	Sunday
				1	2	3
4	5	6	7	8	9	10
11	12	13	14	15	16	17
18	19	20	21	22	23	24
25	26	27	28	29	30	31

How many numbers in this calendar?   30   31

How many two digit numbers in this calendar?   22   23

Fill in the missing numbers from 100 - 71

	99	98			95				91
		88	87			84		82	
80		78		76			73		

Find the sound different and value
(Under line the different)

1	11	21	31
2	12	22	32
3	13	23	33
4	14	24	34
5	15	25	35
6	16	26	36
7	17	27	37
8	18	28	38
9	19	29	39
10	20	30	40

Thirteen → 13
Fourteen → 14
Thirty ← 30
Forty ← 40

1	11	21	31	41	51
2	12	22	32	42	52
3	13	23	33	43	53
4	14	24	34	44	54
5	15	25	35	45	55
6	16	26	36	46	56
7	17	27	37	47	57
8	18	28	38	48	58
9	19	29	39	49	59
10	20	30	40	50	60

Fifteen → 15
Sixteen → 16
Fifty ← 50
Sixty ← 60

1	11	21	31	41	51	61	71	81
2	12	22	32	42	52	62	72	82
3	13	23	33	43	53	63	73	83
4	14	24	34	44	54	64	74	84
5	15	25	35	45	55	65	75	85
6	16	26	36	46	56	66	76	86
7	17	27	37	47	57	67	77	87
8	18	28	38	48	58	68	78	88
9	19	29	39	49	59	69	79	89
10	20	30	40	50	60	70	80	90

Seventeen → 17
Eighteen → 18
Nineteen → 19
Seventy ← 70
Eighty ← 80
Ninety ← 90

Group the numbers and their words

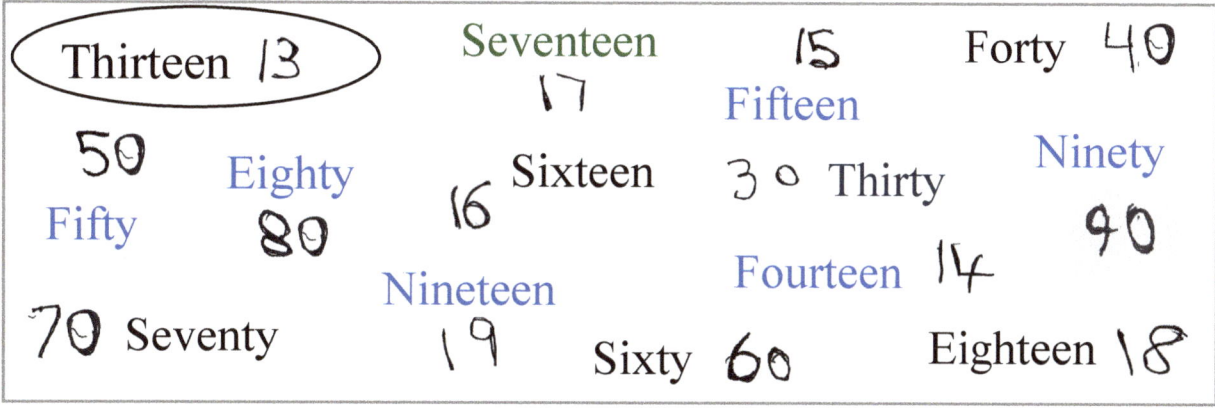

(Thirteen 13) Seventeen 17   15 Fifteen   Forty 40
50 Fifty   Eighty 80   16 Sixteen   30 Thirty   Ninety 90
70 Seventy   Nineteen 19   Sixty 60   Fourteen 14   Eighteen 18

Pictograph in two ways
Count pictures and fill the chart

4	Stars
	Flowers
	Rats

	Buses
	Hearts
	Cups

	Cars
	Birds
	Stars

Learn about ordinal numbers

Circle the picture in the correct position left to right

🚗	🚗	🚗	🚗	🚗	🚗	🚗	🚗	🚗	🚗	First
🏠	🏠	🏠	🏠	🏠	🏠	🏠	🏠	🏠	🏠	Second
🌺	🌺	🌺	🌺	🌺	🌺	🌺	🌺	🌺	🌺	Third
✈	✈	✈	✈	✈	✈	✈	✈	✈	✈	Fourth
👦	👦	👦	👦	👦	👦	👦	👦	👦	👦	Fifth
👧	👧	👧	👧	👧	👧	👧	👧	👧	👧	Sixth
🦋	🦋	🦋	🦋	🦋	🦋	🦋	🦋	🦋	🦋	Seventh
🍍	🍍	🍍	🍍	🍍	🍍	🍍	🍍	🍍	🍍	Eighth
👧	👧	👧	👧	👧	👧	👧	👧	👧	👧	Ninth
🌲	🌲	🌲	🌲	🌲	🌲	🌲	🌲	🌲	🌲	Tenth
1	2	3	4	5	6	7	8	9	10	

Colour the graph in the correct position from down to top

15															
14															
13															
12															
11															
10															
9															
8															
7															
6															
5															
4															
3															
2															
1	▇														
	1st	2nd	3rd	4th	5th	6th	7th	8th	9th	10th	11th	12th	13th	14th	15th

Make the short form of the ordinal numbers and write them in the missing places

1	First	1st

2	Second	
3	Third	
4	Fourth	4th
5	Fifth	

6	Sixth	
7	Seventh	
8	Eighth	8th
9	Ninth	
10	Tenth	10th

11	Eleventh	11th
12	Twelfth	
13	Thirteenth	13th
14	Fourteenth	
15	Fifteenth	15th

16	Sixteenth	
17	Seventeenth	
18	Eighteenth	18th
19	Nineteenth	
20	Twentieth	20th

Underline the different

20 - Twenty	20th - Twentieth
30 - Thirty	30th - Thirtieth
40 - Forty	40th - Fortieth
50 - Fifty	50th - Fiftieth
60 - Sixty	60th - Sixtieth
70 - Seventy	70th - Seventieth
80 - Eighty	80th - Eightieth
90 - Ninety	90th - Ninetieth

Match the ordinal numbers with short form

Twenty eighth	26th
Twenty seventh	30th
Twenty ninth	28th
Twenty sixth	29th
Thirtieth	27th
Thirty third	31st
Thirty second	32nd
Thirty first	33rd

Compose the numbers and write them

Ten	One
□	△

	Tens	Ones	Numbers
1	□□	△△△	2 3
2	□□□	△△△ △△	
3	□	△△△△ △△△△	
4	□□□ □□	△△△△	
5	□□ □	△△	
6	□□□ □□	△△△	
7	□□□ □□□	△	
8	□□□ □□ □	△△△△ △△△	
9	□□□ □□□ □□	△△△ △△△ △△△	
10	□□□ □□□ □□□	△△△△ △△	

(Circle the answers)

What number I am?

1) I am - 3 in ten place and 7 in one place (55, 37)

2) I am - 5 in ten place and 3 in one place (44, 53)

3) I am - 4 in ten place and 5 in one place (45, 13)

4) I am - 7 in ten place and 6 in one place (34, 76)

5) I am - 6 in ten place and 0 in one place (60, 66)

6) I am - 9 in ten place and 1 in one place (91, 19)

7) I am - 8 in ten place and 8 in one place (81, 88)

8) I am - 7 in ten place and 5 in one place (25, 75)

Roman's numbers 1-10   1 = I | 5 = V | 10 = X

1	2	3	4	5	6	7	8	9	10
I	II	III	IV	V	VI	VII	VIII	IX	X

Fill the Roman's numbers in the missing spaces

1	2	3	4	5	6	7	8	9	10
I		III		V			VIII		

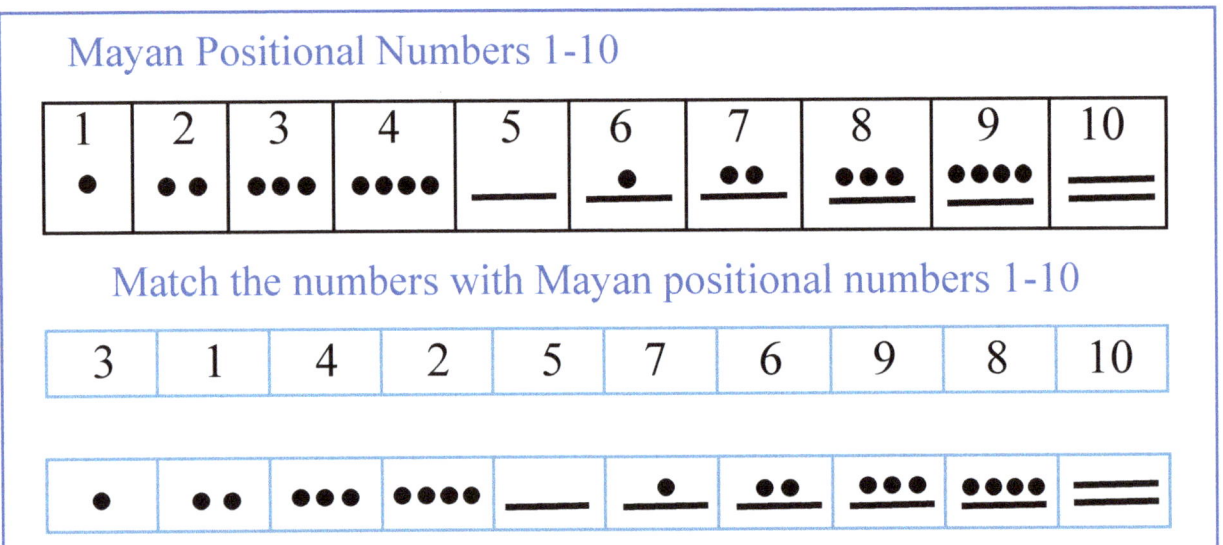

American Sign Language - Numbers 1-10 (Colour them)

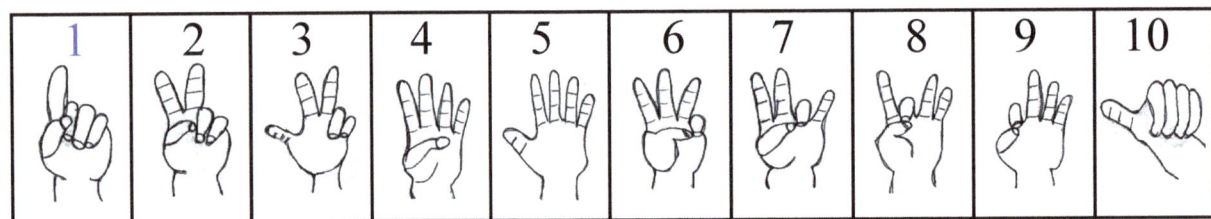

Look at the sign and write numbers

www.ingramcontent.com/pod-product-compliance
Lightning Source LLC
LaVergne TN
LVHW072132060526
838201LV00072B/5017